There's a
THOUGHT
A Conscious Journey

PAUL PHILLIPS

To order additional copies of this book, contact:
Xlibris
UK TFN: 0800 0148620 (Toll Free inside the UK)
UK Local: 02036 956328 (+44 20 3695 6328 from outside the UK)
www.xlibrispublishing.co.uk
Orders@ Xlibrispublishing.co.uk

ISBN: Softcover   978-1-6641-1508-8
        EBook   978-1-6641-1509-5

Print information available on the last page

Rev. date: 03/16/2021

For Debbie my Wife

Q.E.D

The two great questions are how and why
Is anything here, it makes more sense for
Nothing to exist past, present, and future
If there ever was a time of nothing, the
impossible became possible, so in theory
Anything could be!
Possibly say infinite consciousness?
The only end of possible is the possibility
Of not believing

Paul Phillips

CONTENTS

ACKNOWLEDGEMENTS

Firstly, I would like to thank my wife Debbie for her patience while working on this book, my head was quite often somewhere else metaphorically speaking! When in deep thought I often would be distant.

Thank you Antony Price for a re-think and change of name on the book's title, saying conscious was a better word than spiritual, cheers Tony.

Many thanks to Pixabay for the wonderful images I have used in the book, I think they have added a heightened impression on the themes of the chapters.

A big thanks to Natasha Shek for help in compiling the PDF files into one document, I'm not particularly computer literate.

Xlibris staff members Vanessa Diaz operations supervisor, Carlo Dano supervisor publishing sales, Mae Subido control representative and all involved in the production process.

A warm thanks to Chrissie mc'Tagart for our deep and spiritual conversations and encouragement, clearly she has a deep connection with all things spiritual.

Last but not least! All that have given positive feedback and helpful criticism, all of which have given greater insight into my message, and the message is what it's all about.

INTRODUCTION

Today I made the Now decision to turn pen to paper for my next book. The first one was "Its about Time" based on time encounters and our journeys. Although this book does cover time because there is no avoiding it, as we govern our lives by it! This second one is a spiritual journey about consciousness and Human perception, at least a personal one of mine. I look at the depth of Human emotion, to broaden our mindset to more than the superficial. Looking to the future, whether it be technology, biology, Cyber or space travel. I look at an infinite universe where intelligence is more than one organism that randomly appears in a galaxy, but in a whole universe of intelligence! To explore the feelings we have and not to be someones servant. To Be an equal and important part of a collective. How to understand ourselves, and the Unknown! The need to create God to comfort or control or hopefully find an answer to our existence. I express in "Gaia" that the earth is high on peoples concerns and this awareness is a justified concern for our future. Not forgetting the myriad of living things that live here with us and what we depend on. I touch on the side effects of Human intelligence and the complex feelings we have. I tell of my own battle with depression, expressing my feelings through poetry that appears throughout the book. They appear in the beginning of a chapter and form the heart of the message.

They are simple and not cryptic. I admire the great poets and their skill they paste outing there written art, but if you are like me and fall short in deciphering half of them like they are some esoteric society. Ok what is the core message of this book? In a Nutshell, there is no end to possible, only the possibility of not believing, To use The Power of Now! There is so much more to discover, so much more for us to do and achieve. More importantly so much more for us to be! I take a look at space travel and how we might do this, with the great time it would take to travel and the possibility of cryogenics. The colonisation of new worlds through the process of terra-forming, Realising we will have to change and become Cybermen! Not just because it will be physically necessary, but because it will extend and enhance our lives. We will become truly cosmic people that will be able to harness enormous amounts of power, even the power of black holes and nuclear fusion. In two Gaia chapters I make the imperative need to change our ways. What about me? I guess this book has been a long time coming and has grown as I have. All my life I have been a curious fellow staring out of windows trying to make shapes out of the stars. I never really had any real mentors, However when I was at school I was introduced to my first view of the sky through a telescope. Mr Finch who looked like Magnus Pyke the English scientist and TV presenter. He had an old brace refractor telescope, and I was amazed at the Moon with all it craters and Jupiter with its jovian moons. This was a pivotal moment.

In a young boys mental development, I could see for the first time there really was more than what I first could see with just my own eyes! I never did very well at school, in part because I was a somewhat wayward child and coupled with dyslexia. I changed and took to reading more only after I left school. I went down the road of attending spiritual churches, astronomy clubs and gelling with anyone that was a bit like me. Having strange experiences as my life went on, it only reinforced my deep down latent connection with the universal consciousness. This has built up in me on my journey through life, through experience, science, reading, talking and last but not least an opening my Mind! We consciousness on this journey are very much apart of the physical world, living in an expanding universe. We too as conscious spiritual beings too are expanding, taking control of our future through thought and the Power of Now, To soon take control of our mortality and physiology. Becoming more of an introspective species, awakening to the realisation that we have great power and now we are starting to use it. We are restless spirits becoming more aware of a larger connection with another realm, a realm we come from and one we are profoundly apart off. For me I believe we are all from these Gods, these Gods are from our past. It's time we really know what we are for, I think unconsciously we always did. It's time to stop worshipping Man made Gods and work towards being them! We are The Transition to be a God. Throughout my life even as a child I always was aware of something else.

A bit like someone under you're bed feeling. Religious people say they feel this in their connection with God and their faith! This Book is a mix of that and has evolved through my experiences and my little Knowledge, and feel at my time in life I can tell this story as I see it, and that it has been growing in me throughout my life. If we can get past the religious differences, our pugnacious natures and our old perception of us. To be the Gaians that unite to build our future, drawing on what we have in common and putting aside our differences. We have a vested interest in uniting, to be a collective, because deep down we are all consciousness on a journey having an experience of being Human.

CHAPTER 1

The cradle of Man

The cradle of Man

We are the sleepers
That are starting to awake
We are the discriminators
The racists and the preferential

We are the builders
that make the walls to hide behind
We are the architects
that Cradle our minds

The imaginative lines we draw
The constructs of nationhood
Are we not all Humankind
Are you not my brethren my kin

My fathers-father goes back in time
Are you not a relation of me
The Sun rises on everyone
Are we not all Africans

In a mirror reflective
I see a million eyes
A million eyes of consciousness
Diffused into a physical realm

How far do you need to go to find the beginning of Mankind? Thats not an easy task in fact it's an epic challenge. We can trace back some 13.8 billion years with todays current estimates. Science says the universe started with a (Big Bang) everything we see and perceive is the result of a cataclysmic explosion. The universe in its infante stage was compressed and consisted of very basic particles like Hydrogen and Helium. As the Aeons past the stars and galaxies where born and then many of the first stars died many in violent supernovas creating more complex elements like iron and gold and of course Oxygen! It would eventually lead to life. A complexity of chemicals and amino acids form DNA (Deoxyribonucleic acid) forming a double helix made of four types of chemicals bases adenine, guanine, cytosine and thymine. There are about 3 billion bases that make up Human DNA and more than 99 percent are the same as all other people, these DNA's make genes which in turn replicates and mutates in a process of Evolution that connects all life. The story can be found in the rock pages and the fossils within it showing an evolutionary journey that connects all life on earth. A trip to the Zoo should be seen as a visit with our relatives.

Charle Darwin (1809-1882) an English naturalist and geologist conceived the theory of evolution by natural selection where we have ancestors and distant cousins and if we go far enough back we would simply become indistinguishable from bacteria forms. We know that life began at least 3.5 billion years ago because they are the oldest rocks with fossil evidence of life on earth. Let's jump forward say I don't know around 7 million years ago where there were a number of chimp like apes and if you could look at them from our eyes you would be hard pushed to see which would lead to us. Speeding forward again to 2 million years ago someone appeared that you would see a connection with, it was Homo Erectus an upright ape that walk out of Africa and subsequently migrated across the globe.

These new people would be the races we would know today after a few went extinct like our distant cousins the Neanderthals. So from one of those very early apes one of those chimp like beings was going to have a very interesting future, us the Homo sapiens but something also special was evolving and that was our sentient consciousness.

All the races or if you like the subdivided species of Human can all be traced back to Africa (we're all Africans) so thats the physical and biological story of how we came about if not a very basic Anthropology lesson, however its more than superficial than just physical form, there is more to us then just a singular sentient being, I believe we are aspects of a universal consciousness!

When you come to think of it being a racist is ultimately pretty dumb especially through reasoned thought. In fact we are all made of star dust, we are all of the universe for we are the universe evolving.

We are a way for it to know itself, instead of saying we are Humans having experiences in the Universe (think we are the Universe having Human experiences) how we see our world can be simply a point of view. Discrimination is a method of distinction whether good or bad. Race is a very obvious one and did serve an evolutionary purpose it has become more complex today, it's not hard to distinguish between the colours (Black and White). Can you see where I'm going with this? At this point of observation it's just a superficial and visual distinction. It is what you associate with it that gives it Emphasis.

Example

If Men with long hair spat in you're face on more than one occasion you may have a propensity for not liking our long hair spitting assassins and probably have the tendency to socially hang around with people with short hair even effecting a choice in a partner. It would become a natural Bias for short hair.

As we are a social animal we can find so many reasons to discriminate between one another, lets face it since time immemorial the main reasons were for a mate and competition between local groups first in troops of apes then tribes and today Nations!

Sexual discrimination is far closer to home first it was a way of choosing a mate, females would choose strong and fit males and males would be a little less choosy with females but would favour young and curvy figures hour glass shaped large breast, etc.

With the huge amount of migration taking place at for reasons like war, famine and persecution it is only natural to see them as a kind of enemy, today it can be our way of life! An example for instance chimpanzees which attack an outside member of a troop that encroaches on their territory and deep down it is ultimately a survival mechanism. When I see the racist stigma being banded around in my opinion we are being protective of what we have. I find it a load of Hogwash and a bit of a trump card and narrow and narrow mindedness, of-course there are some unadulterated morons out there that are!

There's a pragmatic view to this view point simply a reductive survival and preservation process called evolution. James Lovelock the farther of the Gaia theory, space scientist, inventor and climatologist talks about a balance or equilibrium. It is not natural for millions of Humans or animals for that matter, to be living in such close proximity in extreme conditions like arid and desert environments. This is not conducive of a self-sustainable population and the scales are tipped. We Humans are Victims of our own success through artificial means and inventiveness that we are able to support such high numbers of people, but at some point we will reach a finite state. A analogy what I call the (Lifeboat example) is that there are far too many people scrambling to climb onboard a boat until it capsizes. The simple answer for their survival, if not a bit callous and against the grain of our Perceived Humanity is for some to drown! There is a problem with our over simplified answer we are Human and have empathy and compassion. We have an innate vicarious nature and this is often acted upon. This said can you blame us for worrying if there is a risk to our way of life? Our jobs, availability of food, housing and water, basically good old preservation instinct. It is only Human of us to want to protect what we have, Yes it seems to contradict our empathy side and to a degree maybe a little! But when we are safe we express it as a social animal, however faced with our true ancestral underling mindset our instinct and reasoned thought is to put up barriers and find reasons to repel competition or invasion.

If you look deeper than geographic location, origin, skin colour, how much of you're body you can show, what food you can eat and when for goodness sake what flipping direction you pray. Whether you see yourself as a clever Ape, a creation of God or the intervention of aliens tampering with our DNA we are just the same.

The need for food and water an attraction for a mate the feeling of love and companionship. We have the propensity of exploration to search for answers in the World around us, trying to make sense of it all. Ultimately we want to be Happy and dwell in contentment in just being us! Sound familiar? Of course it does I have just described what it is to be Human at least in my simplistic rudimentary introduction.

Ok did you know that Anatomically our cells are not the same as when we were young, we don't have the same cells we were born with.

Well, not quite true but most of them are replaced. According to researchers the body replaces most of its cells every 7-10 years and some are revamped more often. As we grow old each time they are replaced our cells do not copy quite as well. The body loses grip of its DNA after 55 years, this is the nett result of ageing and looking old.

At this point its worth just reminding you I have not forgotten the spiritual nature of this book but think it is worth getting a foundation on our physical and behavioural nature and I think it is fascinating how we operate and learn in our culture with social insecurities, Though remembering the underlying essence of us. Cells by the Trillions multiplying and dividing

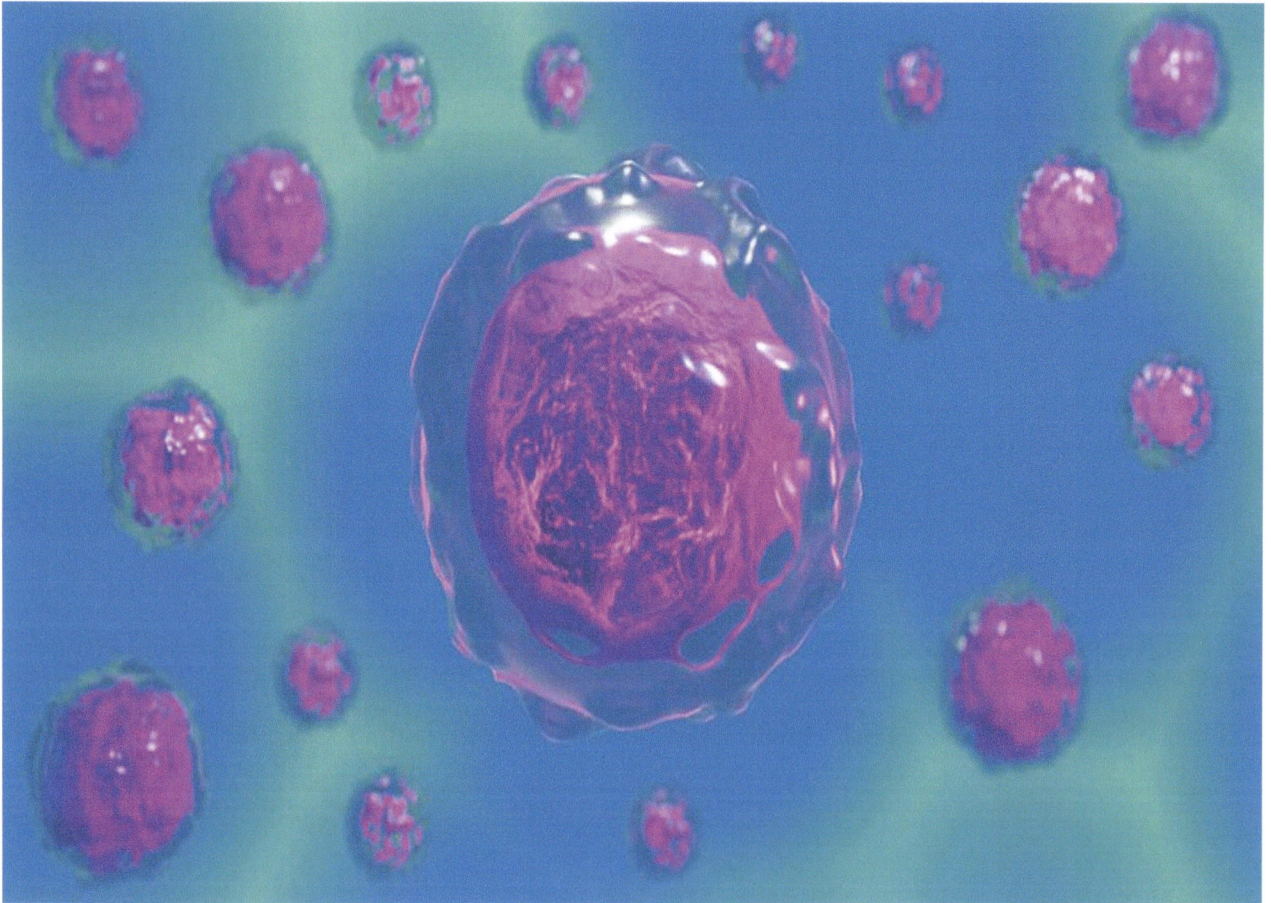

We have between 37-40 Trillion cells in an adult human, some are replaced over different periods of time. Skin cells replace over Two to three weeks whereas brain cells typically last a life time However some do die especially as you get older or have a disease like Alzheimer's or dementia! The marvellous thing about them is they are capable of learning new things, To be well YOU! And importantly able to re-evaluate old beliefs learn by our mistakes, Hey there's hope for me yet!

Ok Brain washing or more technically information imprinting and this is the crux of the matter. The Buddhist faith or as I call it ideology understood and used Mantra or if you like mind programming as a technique, very helpful in learning and taking in information through repetition. Every new thing we learn every new experience we encounter. We're a different person and are in fact we are mostly other people what they say, what they have taught us (We are like sponges) However good at first Sight, it can be an insidious tool.

Radicalisation and promoting extremist behaviour with indoctrination can subsequently lead to extremist militant action ie Terrorism, Now we have come to mind impression printing we are the products of our cultural filtering. With todays insatiable appetite for information and our inherent tendency to worry we can have a recipe for disaster. When you hear about Blacks carrying knives, young men killing each other, we are left with an imprint, mind burning a perception of those men or boys.

We live in quite a violent world and with the problems in the middle east Terrorism is rarely out of the News. Suicide bombers dressed as women in Burka's or just wearing a hijab going about their own business in the high street evokes suspicion and fear of them as they wander by. The connection and saturation of muslims and Terrorism in the media has become endemic in our society.

Blacks carrying knives, muslims are terrorist, what about priests are they all Pedophiles or Homosexuals? Of course not! What about politicians and corruption there has been a lot of hype about MP's expenses and their lies and corruption in their public service. Is it fair to tar them all with the same brush? No, that would be crazy, some of them must be genuine, honest and hard working to represent their constituencies hoping to make a difference.

All this is the outer world of us the human machines, we are distant from our spiritual conscious connection, which is the larger all embracing universal consciousness.

I know I have been Banging on about balance and perspective of who we are and why! but it is important to reiterate the need for equilibrium, although I believe in a cosmic spirit the physical world isn't "Mills and Boon" it's not perfect and is evolving as too is our consciousness, It is growing in us making more of a connection with the earth realm as we are destroying it, we will cover More with "The Tears of Gaia" chapter, Migration and asylum of millions of people across the world is evidence of our Imbalance Looking for a better life with safety, employment and healthcare are just a few examples. The more extreme and intolerable conditions drive the need to flee tyranny and oppression, where there are megalomaniacs hell bent on getting their way, even going to the lengths of genocide. The only way to change all this is spiritual awareness of our deep connected roots, we need a collective prime directive, putting in the mindset that we are all equals and find what unites us. No one job position makes you better then the next, we are all equal we just do different jobs! We have to recognise what we have in common, starting to turn the tide of ignorance and change our ways. We have to help the misguided to put aside their differences.

We can then take on the challenge of our survival as a species and preserve our planet. Money and Power are probably the biggest driving forces in the world today as well as land and natural resources, exacerbating our discord leading to many wars. The plutocrats play Sapiens chess using the populous as the pawns in their acquisition Game. In the future I can envisage a time when there would be no longer the need for money, we would all work to provide, contributing to our communities much like some south American tribes still do to this day. Everyone in the village provides a service, whether it be hunting, cooking, making clothes or bring up the children. They seem quite content living like this.

That said it's a nice thought and won't happen just yet, the first we will see is a cashless Society and at some point money will go!

I wrote a poem called (Between the gutter and the stars) where the desire and acquisition of wealth shouldn't be the driving force of Humanity.

Wealth is a state of mind, conjectural and proportionate to other people, Not what you actually need! If we can put this false Happiness and sometimes a weapon to bed, we could embark on a real adventure, a conscious challenge to put our energy to good use. Remember you are the protagonist in you're own life story, you may be one but loads of ones can make many, You may just be the pied piper For your peers. We are facing a real threat to our planet and we are still not advanced in our capabilities to travel to other worlds to settle. This will take a little time and we need to survive long enough to do so! (We have just one lifeboat at the moment) with exponential rate of rising population growth and the avarice need for more of the worlds Resources. We will one day be cosmic people, However until then we have to look after this one and by doing this we have to start here first with us. Democracy originates from the Greek meaning the (rule of the people) The citizens of ancient Greece would meet in a market space to debate in public topics of the town. This is where the market we have today come's from, They probably had people selling refreshments and other wares.

WE ARE THE WORLD

World collectiveness is the key to our future, only together can we truly move on.

Today we have so many different styles of Government but most have a nominated party voted into office for a fixed Term to represent and make most of the decisions with the opportunity to vote out the incumbent regime. The system of government most of us use is democracy, This democratic archaic and incongruous system is now out tune to the ability of the populous. To express the general consensus through modern formats. As I said the old style was set for say like the British system for voting every 4 years.

In the old days we were ruled by Lords and Monarchy, we lived remotely in comparison to todays traveling abilities, so most rule was from local lords with the Help from the church with the weapon of God to bare! All this was to change in the revolution in the English civil war, Date: 22 Aug 1642 – 3 Sep 1651, King Charles the second, 1649–1651 and his Royalist (cavalier) armies and supporters were locked in battle against the Parliamentarians the (roundheads) lead by Oliver Cromwell 1653-1658, He finally won with his Parliamentarian victory at the Battle of Worcester.

This was a great pivotal moment For our time in history, the common person could have a say in who would represent them, giving them a choice in a party that would at least come close to what they wanted. We live in marvellous age, we are a technological advancing race. We were once subjected to the limitations of speed and great distances: once it would take days to send a letter in our own country and weeks to send and receive one from abroad. The news was painfully slow and out of date in most cases when we finally got it.

Today we no longer have to injure that! We have smartphones, internet formats and of course the good old letter system that travels a lot faster today. We can use this technology to better our freedom of speech and a more interactive democratic system. Understanding that in the not to distant past the wheels of politics were slow, and if we kept having ballot voting on every issue we probably would have not got much done.

Television shows like Big brother, X-factor, Come Dancing and pop idol are evidence that we are capable of gauging the viewers preferences, and within a couple of hours or so are able to choose a winner. Now we are talking about entertainment here and it works. So why can't it work for something as important as a general election, referendums, or wether we interfere in a foreign conflict like the Iraq war. Governments can be quick to give millions to different causes, 650 people in Westminster as in the English parliament making decisions that effect over 60 million citizens, like bailing out banks who have made fortunes out of customers. Take the EU referendum people had enough of unelected people in Brussels passing hundreds of laws for us without us having any say! The time has come for a new revolution in democracy: I shall call it (PIP) People impute policy! In theory the European Union was a nice idea and so would a world government be great, just the kind of collectiveness I have been looking for. Today we are just not ready for it. In the George Orwell's (1903-1950) book animal farm wrote that all animals are equal but some thought they were more equal than others, such as the pigs because they said they were the most intelligent. To see where this could lead if we entered into a global order do please read Orwell's 1984: we have to be ready and have universal Human rights and directives for it to work.

(Regnat populas)
Let the people rule!

CHAPTER 2

Dark clouds and Overcoats

Dark clouds and Overcoats

Dark clouds my sunny days
To hold me in a shadowed embrace
The false persona
That hides the grim
To hide behind the mask
By the actor in me
The rope, the gun
The sweet pill
To end the ache in me
I wish I could be the
Soul of the party
The reaper resides here
Do you really know
How anyone feels
The only one is you
My bed of course my bed
My respite, my escape
To flee the darkness
To go underground
To sleep is to be free
To look for the light
To pierce the clouds
Of how I feel
PP 2014

Sorry for such a depressing poem but thought was important to express how I felt. At first I was ashamed to admit my anxiety and depression, something that has shadowed me all my life. To recognise and accept it is the first step in dealing with it. So what did I do? Well, the first thing and this was the lowest point was to nearly Kill myself. This really shook me. Then I sought help! With the help from my Wife (thanks darling for standing by me) giving me the rock to fight myself out of this dark cloud You're probably be surprised just how many people suffer from it! Globally it is estimated that 4.4 of the global population suffer from depressive disorder and 3.6 from anxiety disorder, Here is an interesting fact that the most depressed country is Afghanistan with one in five people, and the least is Japan with less than 2.5 percent. When you think of the stresses of modern life, the codes, the fashions, peer pressure and the exponential change in technologies. Our societies are subjected to constant and sometimes rapid change, change was a lot slower for our ancestors. For us today in our get there as fast as possible societies the fast pace of our lives is increasingly stressful.

Our media fronted by a familiar news reporter like some photonic member of the family, streaming bad news 24/7 with alarming and sad stories on tap like an endless supply of beer from a barrel. When I went to the doctors and told Him how I felt, my GP said the current consensus was some people have a chemical imbalance. With the Catalyst with Morden stresses or bereavement can lead to suicidal thoughts is increasingly common, I'm an example trust me if you sufferer yourself you are not alone! People with this condition are often prescribed anti-depressants or as I call them Happy Pills lol. There are other forms of treatments like counselling, art therapy, yoga, meditation or even hypnosis. Some may work better for you or a combination of them! I was given medication in the form of (Citalopram). It's a Member of a group of medicines called selective (serotonin).

There are many types of tablets and as we are all different another one may work better for someone else. I can say they have been a great help for me, I felt a bit dizzy coming of them which can be a normal side effect but alas I had to go back on them. Remember when taking anti-depressants they are a drug made of chemicals that can counter the imbalance of chemicals you naturally produce and help you feel better. There are always side effects, I felt sometimes to be a little detached, there are no miracle cures with them at precent. There could be nearly half the people walking around the high street on them. This is not a Be all and end all of the problem. We have to have a perspective view on our lives, a balance between positive and negativeness. To see the good in yourself and of course what you have, building a conscious image of worth. For instance someone may choose to demean or condescend you and this will naturally lower your self-esteem, you don't have to accept it! It's

Only their point of view, You can actually say to yourself you're their equal or even better! You are in fact better than them because you don't demean other people and see them for the Morons they are! Counselling as I said before can help lots of people, having a human being to talk too in confidence can be a great way to review things and hear things you never thought before, a different approach to life. I didn't take this route though! So this is my advice if counselling is a bit heavy for you at least to start with. Then talk to a friend or in my case open up to your partner. A problem shared is a problem halved. If someone cares enough to be there for you, then you have worth. Just one person that misses you when you die, then you're life meet something. Even writing that poem and sharing it has helped and with these words maybe even you! I have noticed there seems to be a correlation and accruing pattern to the problem in creative and sensitive people. Artists, musicians, great thinkers and believe it or not even in comedians. We who are introverted that choose to express our inner and deep feelings through a medium as I have in poetry and this book. We seem to be more receptive and more profoundly effected by our world. Van Gogh saw beauty of life in colours and Beethoven in the depth in musical expression, just two examples of people suffering inside. In a hard and cruel world we tended to be hurt the most. We might not be able to change others and Hey! They have their own problems themselves!

We need to concentrate on ourselves, It could be with some kind of new hobby, volunteer for a good cause or take up a sport maybe Go for long walks or runs and be too tired to dwell on things. All change begins from within. Here are some strategies to help!

6 long-term strategies in coping with anxiety

. Identify and learn to manage your triggers
. Adopt cognitive behavioural therapy (CBT)
. Do a daily or routine meditation
. Try supplements or change your diet
. Keep your body and mind healthy
. Ask your doctor about medications

5 ways help yourself through depression

. Exercise 15-30 minutes a day
. Nurture yourself with good nutrition
. Identify troubles, but not dwell on them
. Express yourself
. Try to notice good things

You're mind is the controller, It's a psycho morphic consciousness able to transfigure our persona to a truly self confident and contended soul. A Happy ego free to enjoy the spiritual journey ahead. Just an adjustment to our thinking we can have a new point of view and a New you is born. (yesterday) I was someone else (today) I am who I choose to be. In many ways we have forgotten that we free spirits can choose.

The worst part about depression
The people who don't Have it
They just don't get it.

Unknown Quote

It's a bit like walking down
A long dark corridor
Never Knowing when the light
Will go on.
Neil Lennon

The only one that truly knows
How you feel is you.
PP

CHAPTER 3

God and all that Jazz

God and all that Jazz

He comes as one as to the many
Fulfilling the void of insecurity
The creator of your peace of Mind
No pondering for me
No proof, no Evidence
Just faith and loyalty
God and all that Jazz
The closed book, the closed mind
All you need is faith from me
There is no need to search for god
He is a figment of your mind
This search for deity
The more I reason, the more
it's not for me

2020 PP.

God now that's a title for a chapter! As you probably gathered I'm a bit of a thinker and admittedly a bit of a dreamer, not that that's a bad thing! I mean just how many things have come about through our dreams. Whenever you have had a discussion or debate on life the universe and everything and yes the answer isn't 42!

And our place in it all, it is inevitable you will consider a creator or intelligent design. Most commonly termed (GOD). I wonder just how many heated debates have Arose on that subject? I wouldn't be surprised if someone has actually killed some in the argument over it! I have claimed for most of my life to be an atheist and in a sense to mainstream opinion and religions I still am. I guess I'm nearer to an agnostic but do have a tendency towards a spiritual answer still not fully understood. I don't believe in a external omnipotent and omnipresent monotheist God judging our every moves or thoughts. This however is not evidence that I'm not a spiritual person! On the contrary I'm gravitated towards a force of collective consciousness, An Intelligence evolving. We are integrated and aspects of a whole.

I am going to ask you to do something that is seemingly impossible to make a point. We can really only relate to the physical or tangible world or what we call reality. I want to ask you to imagine (Nothing) . . . . . Yes, Absolutely nothing, I'll give you a few seconds to try . . . . . . Not easy is it? Bear with me you're get the idea. Now imagine out of nothing the physical world, universe, light, gravity, time and anything you can envisage suddenly sprung out of nowhere into existence. How can this be? How do you get something out of nothing? It doesn't make sense! In fact it makes more sense for there to be nothing at all.

Q.E.D

We exist therefore something else exist's. We are in the process, Everything has a precursor in existence and thus the multi-verse and the infinite possible Dimensions. Does this not contradict a bit with what I'm saying leading to a singular god that I don't believe in? Well, No! I have a problem with what I call man made bespoke religions and this construct of a Heavenly Father. This then begs the question where did God come from? That good old chicken and the egg riddle. Are there other Gods to keep him company in his eternal reign? The statement that he made us in his image is a clear example of mans vanity. A lack of imagination using us as a reference! Are we so pretentious to say he created us in his image? Is this the extent of our vision? Let's try to get our heads around this one, something that has bugged me every time I think it. If in the beginning there was god and he is eternal he's been around a flipping long time, plus if he is just one then we can presume he was on his own all this time, suddenly out of the blue creates us all out of that unfathomable amount of time. I mean what took him so long? And man in his image Please: would he make such an imperfect representation of himself! So after this time did he just make a pet to keep him company?

Just imagine a sole God or entity being alone for such an unimaginably length of time wouldn't it make you a little bit mad?

Just a reminder for simplicity I use the term God as most people have that in their minds. What is God! The life and force underneath and in all forms. What is love! Love is what we are deep down within, it connects us like a force of attraction, therefore permeates all forms. Thus it's the state of God. This can manifest in different leaves and degrees. The problem with man made religions is they make a simple explanation into a more complex one with "all that jazz around it" namely customs and rituals. I am not saying I fully understand the great depth of that power I think you have to be closer to understand anything. This is what I think is best not to spend too much time trying to quantify it. Don't dedicate too much of you're energy, instead just feel and become immersive. I will add more in the Chapter (Evolving-consciousness). As I have said about the multi-verse. I believe there is so much more going on in a possible infinite and endless multi-verse, so many dimensions it would be impossible to know them or relate to them all, especially from one unique perspective. Only the ones adjacent or interlocking would be able to notice or perceive the presents of another realm of existent. As our universe evolves, and indeed us as a part of it will one day know the next. I am doing my best to put a distant from that old image of a bearded God looking down on us from some heavenly throne ready to shoot a bolt of lightning the moment we lose favour. The Un-manifested is the spirit of presence that underlies and radiates in us all, when I say all I mean all! An ant to a whale, tree to a sea cucumber, in all living things. As to inanimate things a grain of sand to a mountain. This God is in it. Stop for a moment and think: yes! It's in you're mind too. See it like electricity and volts in a cable. We have more of this God than say an ant and that is because we are able to handle more than an ant, the ant more than a stone. We humans are more connected due to the power of our minds, that old cliche love is all around us is only partly correct. In fact it is all centric in us and to a degree the most minuscule particle in existents.

I shall be repeating in more detail this next bit but is central to the ever centric relationship. Close you're eyes and say to yourself what will my mind think next! When you do this an aspect of God is watching you're mind and form. You are not just you're mind, just as you are not a mountain! If you were just you're mind which perceives the outward and your body form, it wouldn't be able to perceive itself. It just can't it needs the observer being at the wheel. I am writing about my mind . . . . I Am in my mind. Now you will know you're part of a larger consciousness to which you are an aspect. You are not a mind having a conscious experience but instead you are having a worldly physical manifestation through you're mind experienced by your consciousness. When you asks yourself what will my mind think and tried this a few times you will start to feel you are more than you're mind. When you have true communion and connectedness you will realize our oneness. This realisation of oneness will one day end all wars, poverty, discrimination and the drive for wealth, The subsequent ending of the rule of the plutocrat's the brake on humanity will be released. We would then be able to see our true course that of love and the joy of our being! A realisation of being Part of an amazing wonderful multi-verse, therefore a closer love for us in the Majestic story of Everything.

Now a little science as we have now become accustomed to it. To answer our inquiries and as an alternative to religion. Life is difficult to grasp because we don't understand it. It's as if living was some kind of sin. We are born without knowledge of where we came from at least no preprogrammed reference point, living short lives, feeling pain and losing love ones. We go on our journeys through life getting older, losing our looks and with that good old Gift the knowledge of our mortality. So often we are bombarded with religions and cultural norms from dictatorial powers! All of them haven't been able to offer good evidence to their beliefs. Telling Us to believe and follow and practice the faith! Or God help you you're get it! Today we live in a more secular world with science offering an alternative physical process of how the universe and life came into being. It is at this moment not the definitive explanation, but it has helped dispel many inaccuracies in the frame work of religious thinking.

I do feel we are on the right track taking into account our ultimate spiritual essence. For name sake I will say we came from God as its a term everyone uses but in fact its a spiritual collective intelligence. For us we are Gods in our infancy, Gods evolving. We will do God like things and one day there will be someone praying to us! In Buddhism it is said that the more you search for yourself the more you Will not be able to find it. The more you look at you're body the more unrecognisable it becomes, becoming fragmented or quantum. Today with our understanding of physics we know this has been proven to be true.

The universe as we understand it is made of countless Atoms intern made from sub-atomic particles. These small elementary pieces of matter were first conceived to exist in ancient Greece and China. The name Atom comes from the latin Greek (Atomos). Democritus held that everything was composed of Atoms which are physical. Between them Is empty space, in part correct but we now know that inside them is a nucleus made of sub-atomic particles.

Atoms are made of up of three kinds of smaller particles called Protons, which are positively charged, Neutrons which have no charge and Electrons Which are negatively charged. They can be in different numbers as they orbit inside and around each other. Inside them we go smaller again and find Quarks and Gluons. There is a lot going on in this sub-atomic world and a lot more sub-atomic particles I have not mentioned!

Needless to say we are made up of a sub-atomic soup. While this soup is bubbling and moving there is something also going on. (Energy) there is no getting away from it! It is said you cannot make more of it in this universe, and you cannot destroy it. Referring Back to Atoms I aforementioned positive and negative charge. Does this sound somewhat familiar in day to day life? God good or positive, the devil evil or negative. Light and dark, positive thinking or negative thinking. This is the ingredients of life, every time you stop and look for this pattern you're find it. No wonder it appears as such a key theme in religions! Let me ask you if you are a religious person of any faith not to think of you're in-doctrine! Do you know right from wrong? Do you know that you love you're family, husband or wife or even you're pet dog? I know I love my wife and don't have to read a Bible to know that! What about how we behave with strangers? IF you witness a road traffic accident of total strangers do you need a religious view point to feel compassion. This is innate its part of our nature! I believe that the concept of God and the systems people follow are in essence the same thing, but without true understanding. We can't ague some of the core values and yes there must have been something before us that created us. Are people good and bad, Yes of course. Can you argue With Thou shall not kill, steal or commit adultery. These core values we have in Human society are not the foundations of proof of God, and the validity of any particular religion.

The problem is the way we interpret our faiths and thus practices we follow. This is where it goes wrong, we can't deny life begs an answer and we do feel insecure. May I say that most do generally want to live fulfilling and meaningful lives, we are at heart mostly decent beings! The problem as I said is there are those who use it as an excuse to encourage people to act in extreme ways. Let's look at this in a little more detail. Say you're a devout follower of you're selected religion and want to do all that is possible to show that you are, then you're going to do as you're religious leaders say. Ok a none believe may still commit a murder with the knowledge of a life sentence or execution but that is it. A devout believer will believe in condemnation to eternal punishment in hell! They also depending on the teaching it may also happen if they don't do it in the name of God, Throw in the thought that you will enter paradise in heaven if you do, then you will be able to get people to do anything! I mean in their eyes they have nothing to lose in this life because a better one is guaranteed in the heavenly after life. God the all powerful and judge that created the universe is seen as the arbiter of all our sins. In my opinion if God in the traditional perception is true! Then it is his job to dish out the sentence not us. I mean the arrogance of man to act in the name of God without understanding the mind of Him, her or it?

CHAPTER 4

As if living was a sin

As if living was a sin

What a choice, we have no choice
We're born without a say
First light the world is bright
We know not this place
Like a bio drive
We await our programs
To learn the world we find ourselves
Vulnerable and reliant
We seek parental guidance
A nurturing tender hand
Chemicals with a conscience
Our bones with a mind
Taking in the panorama of life
We endeavour to make the sense
Religious and secular
Having a plethora of choice
We meet the conundrums, the paradoxical,
the multitude of paradigms
The universe has many centres
My mind is the centre of mine
We know our finite time
To spend in this place
All is out of our control
to an uncertain end
No autonomy in being human
We're bound to this place
As if living was a sin

I can't understand for the life of me
the torment the injustice
the dictatorial existence
The power over me
To add insult to injury
We know not where we go
Does this sentient being
have an absolute end,
Does this soul evolve and grow
To find the truth
Is when we go!
PP

Well, how do you comment on that poem? It's fair to say our lives are one hell of a mystery. We are born blank like some kind of bio computer awaiting to be programmed or should say in most cases indoctrinated. A self-learning intelligence device called the brain! Perceptive and responsive to our environment, whether it be observing a sunrise or cross legged listening to a spiritual leader, but for most of us it is a teacher at school passing on indoctrination. Each generation is a recipient of their time and understanding! We are victims of culture filters and generally act accordingly. Knowledge is a kind of evolution mutating slowly or exploding rapidly to give us a new view of where we are in this universe. Religions are much slower to handle the change! Over thousands of years we have built up a humungous mountain of information on our world and are spoilt for choice to what to believe in, There are more religions than hot dinners. All of them are Human constructs. Those who believe will apparently find paradise: God if you pardon the pun: for Me, I have no chance! We are told he is a loving God and he moves in mysterious ways. Stop! You mean we don't understand him is what they are really Saying! All this choice in a multitude of faiths and customs, The biodegradable beings we are and the mental stress attributed to our awareness, plus our perceptive minds we are so conscious. The flaw if you can call it that? Is we question everything! The biggest enemy of the traditional God is the age of reason. We are just expected to believe.

Just have faith Man, follow the good book and every rule! My name is Paul not fool, Sorry if you don't mind I won't get onto that wagon. For me I just think it is a manifestation off I don't know, I need a heavenly father to look after me, am I just a mortal? It draws on comfort and hope. Life is easy when it is simple, so I don't have to think and of course any answer is better than none! Even if it is wrong. The route of all unhappiness is not having what you want or doing as you like. Take the apple in the garden of eden, a nice ripe fruit juicy and good to eat. It seems we are in a battle against our inner desires. To make love, No you must be married first! You can't eat pork but its ok to eat beef? Don't show the natural beauty of you're face if you're a women. The big one for me is if you are not in a faith then you must be soulless, evil and sinner! Or how dare we think we are apes, clever ones, but nonetheless apes. We are aspects of the deeper universe here to have an evolving experience and what you believe doesn't make you bad in most cases, It only can make you wrong? Enjoy that bottle of wine, that roast pork that free love making as long as it "doesn't hurt anyone" Believe in what you won't and take it to the grave, let God judge if thats what will be. Man can step back and let you be free.

(For to be free is not merely to cast off one's chains, but to live in a way that respects and enhances the freedom of others)

Nelson Mandela

CHAPTER 5

Paranormal

Soul Journey

Sunshine, wind and rain
A soul enters the world again
A baby is born, and it's very first cry
Born to our world and destined to die

Born to die our soul will fly
As our spirit rises to the sky
Floating Out of our bodies
Leaving people we love to cry

Soul Journey to the astral plain
A life force that needs no brain
A world of no material bound
Where there is no hate and only
Love can be found

There was a world before us
A world we don't know
Traveling through this one
To another we go.

pp

When it comes to events outside our understanding, strange unexplained phenomena or strange happenings we often categorise them as paranormal events. Because we can't explain them we either dismiss them giving them a man made or divine explanation. If this is not used than the experiencer is either on drugs or a Lunatic! Now I'm not saying every UFO "Unidentified flying object" is an extraterrestrial spacecraft looking for a Human being in some remote location to encounter, that no one will ever believe to anally probe them. And whats that all about? A highly intelligent species travelling numerous light years to look up our backsides! I mean what do they think they're going to find up there the meaning of life or a portal to another dimension?

Thousands of people have claimed to of seen ghosts, ufos, Bigfoot and the Loch Ness Monster to the virgin Mary. In most case's they're not what they first initially thought they were and have perfectly reasonable explanations. However we live in a large and varied universe, seeing in light, which is only a fraction of the electromagnetic spectrum. All the stars and galaxies we can see is only about 4 percent of this universe. The other is 96 percent of dark matter and Basically I don't know! Everything in the universe has a frequency. All matter vibrates on different frequency bands like radio or television. When watching one channel you don't see the others but that doesn't mean that they don't exist. When we start to delve deeper into matter, as in subatomic particles and energy we enter A weird world of quantum physics.

A world where an atom can be in two different places at the same time and time itself is relative and not absolute to the speed you are traveling. The more closer you look into matter the more space you find between the particles. We do indeed frequent a strange universe. Astrophysicists say that things only exist if they're being observed and to make sense of everything is there to be other universes or dimensions. The philosopher George Berkeley 1685-1753 said that physical objects that exist do not exist independently from the mind that perceives it, physics says a similar thing. You may say that if you see the moon because you are looking at it and I don't because I'm not looking it, that doesn't mean it's not there! Partly true for me it's not because my world can only be experienced in my mind. Remember it doesn't matter who is looking and who is not, the fact we are universal consciousness means it exists as a whole. The past, present and future happening at the same time, Our consciousness is of all time the physical world is quantum events the true spirit observes.

Let's take a look at the word (Paranormal) the title of this chapter. The dictionary says it means beyond scientific explanation. The para means beyond, apart, beyond and abnormal. The second part normal, well normal! When the supernatural same meaning or paranormal are just that until scientific discoveries and explanations class them as normal. What was once beyond will become understood the next".

As we become more conscious and our minds become more sensitive as it grows and evolves we become more aware of our true connection and Origin. As I mentioned before if scientists are correct and the universe and all events past, present and future are all happening at the sometime? Then so too does our consciousness even if we are not aware of it. The short stories I am going to tell will make some sense! I hope so! These are some way connected to me. The first one is my own and the next ones are my friends.

A Book on roses

I used to work in the London borough of ealing in grounds maintenance. We had a seasonal worker called Mo" who wanted to gain full time employment, a likeable lad who under the usual and old way would have been offered the job. The current manager was fed up with none experienced temps being unreliable or just not that interested in this kind of work. He was looking for experienced or qualified candidates! I said to Mo" to make out he was Studying Horticulture and to get a couple of books on the subject, a folder and note paper that kind of thing. We were working in Walpole park ealing and I knew there was an Oxfam book shop walking distance from where we were working. We decided on our dinner break to go to the shop. It was quite handy as I had been thinking for sometime I would like to get a book on roses myself. So we went in and quickly found the shelves on hobbies, birds, wildlife and gardening! We found a book on general gardening for him and saw a couple of books on roses. Both written by the same Author, Dr D. G. Hessayon, Mo picked up his and then I picked mine. Now remember there is a fifty/fifty chance of which one I was going to get! To my surprise on opening the book my surname was written in biro on the first page, spooky or what! Yes, you could say it was a coincidence or you could say a coinciding event?

Tracy young

Again in the London borough of ealing in a town called Greenford. I went to visit my friends, on arrival I was made an obligatory cup of tea" and we retired to the front room to chat. My friend Bryan was excited to tell me about a strange experience he had recently had. Bryan is older than me and favours music from the 1950s-60s, I do like the music from those's time periods but it's more his age group than mine. He told me he had a dream about Madonna like most hot blooded males do LOL! Seriously she musically was not his cup of tea. So no real connection for him as could be an avid fan! The next day when his daughter came home she had purchased a CD of Madonna and blow me" it was produced by none other than Tracy Young the name of his daughter in full name rather than just like my surname in the previous story. While writing this chapter I had another weird experience. I was looking through a website looking for free download images to use in my book early before going to work and was thinking I had to speak to my Director about a couple of things, The website boasted 1.8 million free images, when opening the first page the welcome page had random pictures of scenery, animals, and buildings that kind of thing. To my surprise there was a photo of a couple sitting on a bench and the man was a dead ringer for my director! I sent him the photo to see if it was him, he said not. I know they say we all have a double out there. I have seen mine an Asian doctor in bridgwater Somerset. Incidentally I have not been able to find the image of the couple again, Strange but true! There have been so many unexplained events in peoples lives! This next account hits accord with me because if it is true would support my idea about physics and the spiritual realm, an inseparable connection. In fact we are not truly separate as normally thought but extensions of it. Energy consciousness vibrating at a slower rate, the (material). This next account is by a neurosurgeon Dr Eben Alexander. Suffering from a rare bacterial meningitis fell into a coma and claimed to of had an NDE or near death experience. This Harvard Doctor of 15 years considered himself a man of science and like most men of science was inclined to accept empirical evidence, but this was going to change his mind with this experience. The condition disables all cognitive brain activity meaning you will have no thoughts, dreams or any kind of imagery. In his experience he said he meet his spirit guide and seeing beautiful landscapes, traveling through deeper into a different realm

What is interesting to me is the description of a sense of there being no particular time and of being multi present? A oneness with it all, and an awareness of higher levels of existences. To me it sounds like a transition from what is familiar, landscapes etc to something much deeper normally out of our range. Quantum scientists say for our universe to make sense we should included a concept of Muti-verses.

Ghosts

Since time and immemorial people have claimed to of seen or experienced ghosts, probably the most common and numerous paranormal encounters of us Humans. I bet you would be hard pushed to either of not had one yourself or at least know someone who says they have.

My Encounter

It was in the mid 1980s in the summer time in Northolt west London middlesex, it was a hot sticky night. Remembering back I had not been drinking and was of good health. I awoke feeling a chill and discovered my cover was not on me and it had fallen onto the floor beside me, well I hope that was what happened! I pulled it back over me and was suddenly aware of a presence in my room! I saw a Human figure standing there at the end of my bed slightly to the side. I thought at first sight it was my brother but within a second or two I quickly realised it wasn't.

It was tall for him and judging by the size I took it to be a man. My bedroom was up stairs and we had dogs! I fast realised this was not normal. My visitor was standing and what looked like it was looking past me judging by the angle. Remember this was in a very short period of observation time. I had a sense that I had interrupted something of it's time, something it normally did. I was visiting it rather than it of me? It was at this point of time that I became very scared: This is the point you can laugh I did the most manly thing you can do in that situation, No I didn't try to make contact and communicate to ask who they were. I just pulled my cover over my head and waited. Feeling a sense of presence and half expecting my cover to be pulled off me, but nothing happened. I don't remember falling asleep as I was too disturbed, I only emerged from my cover when I heard my younger brother Keith get up for his paper round. Calling Out to him he entered the room and turn on the light. I proceeded to tell him of my experience! The strange thing was the length of time till the morning it seemed to suddenly arrive. I never did see it again but for sometime was very much aware of the possibility it would return, so real to me it felt.

"Lieutenant Arthur the Ghost"

This story has a twist and coincidence and goes like this... My old school friend Kevin's father named Arthur served part of the second world war guarding on an air base in Montrose in Scotland. He describes the nights events as followed...

It was a moonlight night and he with another soldier was on guard duty patrolling the base when they saw an unidentified person on site and promptly challenged them by calling to them to stop and identify. They took no notice and didn't comply so they fired a warning shoot to no compliance! Arthur then fired at them thinking it was a German and Knowing there were POW'S in the area. He told me that he was a good shot and with a few well aimed shoots didn't drop them! Both soldiers panicked and ran to get help, by the this time the airbase alarm went off and subsequently the personal became active. When the CO, Commanding officer asked for their report they commenced to tell him. After they had given their account he replied you have seen the Ghost of "Lieutenant Arthur" and it had been seen before. I thought he might of Been joking with me as the name was the same as his.

Some years later

Once again working in the London borough of ealing I had gone into the office to report for work and noticed the Acton supervisor had a book entitled (haunted airfields) by Christopher Huff. I was intrigued and thought that looks interesting and asked could I borrow it? When he had finished and he said I could take it now. When I was on my dinner break I started to flick through the pages and looking at the chapter names, to my surprise saw a chapter on Montrose airbase, so read with interest!

To discover there was such a person called Lieutenant Arthur 1884-1913. He was a RAF pilot and around 7.30am on Tuesday 27 May 1913 was flying his biplane when it started to lose altitude and descended from 2'500 feet when his right wing snapped off Plunging to his death. His Ghost is said to haunt the airbase and has been seen on many occasions, today over a 100 years later people still report sighting it.

Coincidences

As you see there seems to be a lot of coincidences going on in my life, and do you know what? This is very common for me! So often I just take it as normal. Just this morning My wife was getting ready to go and get her hair done at the hairdressers and I noticed a lot of washing up in the kitchen bowl. I don't do much washing up as Debbie tells me not too, as she says I don't do it properly! I can't remember the last time I have done it. I thought as a nice gesture I would do it for us so when she comes home it would be tidy. You might of guessed? She said to me if I do it can I do it properly! Now how did she know I was going to do it without saying so?

UFO'S

An unidentified flying Object (UFO) is any aerial phenomenon that can't be explained most generally on investigation subsequently are, most are in fact satellites, meteors, clouds, Venus or military aircraft. I have been interested in Astronomy for most of my life and I can't say for sure that every dot of light that moves in the sky is man made, atmospheric related or astronomical. But most are! In the 1940s fighter pilots would chase Foo fighters, by the 1950s with the advent of space flight the public fascination in all things extraterrestrial increased, so did the reports of fly saucers. Let's just keep a level head here and see a pattern in more mistaken sightings and over imagined fake reports. Not all are I believe and you will see the link with spirits, NDE'S, and possible extraterrestrial encounters.

Kenneth Arnold

Kenneth Albert Arnold 1915-1984 was an American aviator and Businessman. He was the first to coin the term flying saucers, in fact he said the objects flew like they were skipping on water like a saucer would. He was flying over Chehalis Washington to yakima Washington when he claimed he saw 9 shiny unidentified flying objects flying past mount rainier. On trying to work out what he had seen he had ruled out sun glare on the glass and birds. Estimating the speed of 1,200 miles an hour, this was quite a speed for the time! When Arnolds story got to the press the flying saucer gained the popular term for UFO'S.

There have been some big stories since then, like the famous Roswell New Mexico crash, spaceship or air balloon? Rendlesham forest Encounter in Suffolk England in 1980s just two examples!

Betty and Barney Hill

Betty and Barney Hill claimed to have been abducted by Aliens. They were traveling on a rural road in the state of New Hampshire USA on 19-20 September 1961 when they spotted a bright point of light in the sky moving below the Moon and Jupiter, after driving for sometime observing it they stopped to view it more closer through binoculars. Barney carrying a Gun! Stepped away from the car and he could see the occupants 8-11 he said of them. A long structure descended and they fled away, while they did so there was a buzzing sound and a tingling sensation. Wait! It gets more interesting! They said they had an altered state of consciousness and then returned to travel the next 35 miles, only having a vague memory of it. The journey should have taken 4 hours, they arrived 7 hours later. When asked about this they had no explanation. The first story of Arnold is more believable as these could of Been aircraft of experimental or top secret planes with the hight and brightness confusing the image. The next one with Betty and Barney is much more out there! Either they were Lying or it did happen? What is of great interest to me is the missing "Time" aspect, a pattern that keeps appearing...

Making a connection

Close Encounter of the first kind

I will now try to make a connection with visitations and out of body and Near death experiences. As I have said before most of the UFO's Are far more common as distant sightings as lights in the sky, shiny air balloons, birds and atmospheric effects. All can be mistaken. I think it is important to keep a level head and perspective on Human encounters, they may well of had one but not always as they seem!

Close Encounter of the second kind

A UFO seen and interacts with the environment, this can be electronic interference, Animal reacting, Some kind of form of physical trace. Many sightings have been supported by Radar. Crop circles, burn marks and imprints. Sound and video has now become more prevalent with video recorders and smart phones. We now have some evidence other than personal claims.

Close Encounters of the third kind

This Encounter is least common or believed for that matter, A UFO in which an animated sentient entity (Alien) are seen, robots interacting with the environment or the observer and experiencer. We are now entering a realm of either incredible experience or a great fib! If they are not lying because this is so much a closer and physical encounter than this should be seen very much more seriously.

Close Encounter of the forth kind

Often put in the third category this encounter is even less more publicly common to be reported. It often gets put in the remit of mental health and I just think that can be just a little unfair in some cases! This is where my connection comes in. We enter the world of abductions that take place with one or more people. Sometimes people only give accounts when under going regression sessions and Hypnosis with a psychologist, psycho-therapist or even Parapsychologist who would be more experienced in all things paranormal.

Deep connection

This is the point where consciousness meets the apparent tangible world. We have all kinds of descriptions of these visitations. Human spirits are often thought which of course maybe possibly true? In the classical sense but I want to go deeper than this. We could just be looking at a lot of different events that maybe the same-thing? Every Visual, Smell, Taste, Sound and Touch are the mechanics of our bodies but ultimately we perceive them in our minds, you experience all of this reality through our minds and observed by our consciousness. If Ghosts are real and cosmic visitors to, then it stands to reason they have to come from somewhere? All life is really energy in different states of speed, low vibration more material, High vibration more energy. The more higher the vibration the more none solid and physical. By my definition more conscious energy! We are conscious spirits having human experiences. Thus this is why some of us in this evolving mind have these experiences. Why can they be so relatable? Because we are still very much connected to this realm so a bit like a toe in the water. Dreams are real because we are energy and this is us, because our minds are real, our Our consciousness is real is real.

Rubin

As you may have guessed by now by having strange experiences has kindled my interest in the paranormal and deep down I think that we all have an interest to some degree. This is another account of my strange coincidences. Back in the late 1980s or very early 90s I never did log the exact time. I used to go to spiritual churches with friends. On my first Visit to one I went with a few friends to a small church in Southall west London.

A small corrugated building and of all the colours it could be, it was painted pink! In a small plot of land covered with overgrown vegetation. When we entered there was a tall gaunt man dressed in a black suite and looked just like an undertaker! Moving Further into the church and to the left of the centre aisle from where we were going to sit was a young woman sitting with a hooded cape with a very pale and pasty complexion. I felt a bit of a Berk, sitting there, it felt so weird! I was clearly out of my comfort zone. Remembering the night well it was winter and the weather was bad with rain with a strong wind. Throughout the meeting the wind was howling and the vegetation by the church was rubbing and scratching the sides of it. Retrospectively it was like something from a horror movie. Being young and with friends I thought it was a bit of a laugh at the very least, however I did have a pre-perceived notion I was going to see a full floating apparition but alas they didn't appear. When the spiritualist went from person to person I was impressed by how accurate he seemed to be. When he came to our group he picked Bryan first and said do you have anything to do with boats! At that point I burst into laughter. Bryan was wearing a white woollen jumper and unusually was adorned with a beard, so looked a bit like captain birds eye! My laughter was a bit infectious but we soon composed and settled ourselves. Then last but not least came to me. I was asked do I teach? My reply was NO! He said I was going too. I thought this a little odd as I have dyslexia! With hindsight as you can see It is possible with some effort, anyone can! I Would do so with my Horticultural work with inexperienced staff. He then said he was aware of the presence of a man called Rubin and asked could I take that? I had no idea who it was so NO! A very usual name, he said he was either family or very close to it. After the meeting I left a bit confused and sceptical, as time went by it played on my mind that unusual name Rubin! Who was he? If true someone must know who he is! I asked my mother and again NO! When I saw my Auntie Dolly then in her early 90s was surprised to hear me asking about him as she had not thought of him for sometime. She asked why I have asked her? So I proceeded in telling her about the spiritual church and medium, looking oddly and with curiosity she said he was a very close friend of the family long ago and was always around, they saw him as one of the family.

Paranormal facts

Given the handful of personal accounts in my life and remember there are thousands of other peoples experiences, so not unique to me! This is also not unique to this time period, it has been going on for and spoken about throughout history. The bible is a good example! Somewhere right now as I write someone is having a paranormal experience. I bet you can or someone you know can tell a similar yarn! How many people have seen a UFO? Well, not an easy question to answer. I will use the united states of America as an example, One because it has a population of over 325 million people and two because of the ease of reporting. It is estimated 1-10 people actually report them. For 10 years from 2008-2017 the combination of the NUFOC and MUFO data gives an average of 10,840 sightings a year. A fox picture pole gives a higher number indicating 16.74 percent of Americans say they have had a UFO sighting. If you take into account the whole world, plus the unreported cases it is a considerably number, in the hundreds of thousands! May I say that it can't just be whisky swigging and acid popping junkies who claim to see them, many rational and professional minded people also see them? Like airline pilots, Doctors, police officers, teachers and priest's to name a few. Can all these people be mad, lying or hoaxers? NO, I don't believe that! A lot of these people have important careers and reputations to uphold.

NDE'S

Surveys conducted in the USA, Australia and Germany suggest 4-15 percent of the population have had Near death experience or something that gives them the belief they have? Taking into account that is only 3 out of 195 country's it could be quite a tidy sum.

Coinciding events

We often call these just coincidences and the number of them must be astronomical? I think I can confidently say that every person that lives or has lived has had at least one strange event like thats uncanny or Déjà vu! What is going on? Why are so many people having these strange experiences? Myself included! I don't see myself as Mad which probably means I am, LOL! A little eccentric maybe, I would like to think I am an open minded objective kind of person not given to just believing straight away in anything. If anything I'm like my mother quite sceptical and pessimistic by nature. Not in every sense is it a bad thing, in fact it can make you grounded and more analytical, stopping you committing by just feelings. How common is it when telling someone of a mysterious event you or someone you know of get the response like, I would have to see it to believe it, I have never heard of that before, it must be GODS doing or have you been drinking? Throughout history any new scientific concept or invention is Met with a pre-perceived sense of wizardry or even tampering with God's creation. A spiritualist, scientist, or pagan herbalist could of been sent to the fire at the stake. Visiting comets like Halley's comet named after the astronomer who discovered it, Edmond Halley 1656-1741. In fact it had been seen before but he predicted its return through the laws of motion. They were once feared as harbingers of doom. Let's think about eclipse's, imagine the fear of a full solar eclipse turning day into night: without astronomical knowledge how scary was that? Benjamin Franklin 1706-1790 once seen as the wizard of electricity is seen today as a pioneer of electromagnetism and physics, a scientist. Today we enjoy all the benefits of those pioneers that took the brave steps forward often unwelcome by cynical peers or public opinions. If I asked most people today do we orbit the sun? The answer would be yes! But for poor old Giordano Bruno 1548-1600 an Italian philosopher, mathematician, poet and cosmological theorist, He believed there were other suns with planets orbiting them like ours long before we knew this! For the church everything revolved around the earth. If the church could be wrong about that, what else could there be? To risky only one thing for it? While he insisted the universe was infinite, sounds familiar! What is interesting and pertinent to this book is he believed and was an advocate for the transmigration of the Soul and reincarnation, This didn't go down well in the eyes of the church and the inquisition found him guilty of heresy and burnt him at the stake!

Dreams - OBE'S

This is an account of mine that took place in or around 1989, I had what you could say an out of the body experience. I am not saying it definitely was because dreams can be very vivid? I went to bed as normal with no influence of alcohol and was of good health. As I remember not stressed or anything playing on my mind, I went off to sleep relaxed. I will call it a dream for all intents and purposes may very well of Been? In this dream I felt I was floating and lifted out of my bed and moving around the room! It was a wonderful feeling, Was so excited. While I was doing this I could sense the presence of others but never saw or heard them. I do recall showing off that I could fly! Finding myself outside the house ascending high, this did feel somewhat disconcerting, I mean hell, We just don't do that! I was afraid I was going to fall. Exhilarated as the experience went on I found myself traveling over landscapes like others have claimed, I didn't recognise the scenery. The funny thing is I was not unaware of navigating the journey I just seemed to be doing it like drifting down a flowing river. It became blurry in time and then I awoke. This is my room I mean it really is my room! I could see the room like normal but this time it seemed new or more vivid. I felt so heavy like there was pressure weighing down on me, When trying to get on my feet. I can only describe it as if for the first time I was really feeling gravity. This subsided after a few moments.

On reflection I was sad I could no longer float as if some wonderful gift had been taken away from me. Why do we sleep? Scientists don't really know but it's thought that it gives the body a chance to rest and regenerate. In this state the brains energy can concentrate on working on you're metabolism, generating vitamins and aiding body repairs like cuts. As we are a conscious creature with a lot of information processing going on. I have coined the term (IDF) Information defragment filing: Think of you're brain as a bio-computer in sleep mode, you shut down and rearrange millions of bits of Data you have taken onboard either conscious or unconsciously. This could be anything from an important meeting like a first date with a pretty girl or good looking chap! You wouldn't want to forget that would you? Apart from that you also take in loads of what I call Background accumulated clutter (BAC). We really do need sleep. We become very vulnerable when we sleep, with predators, so we must have a very good reason to do it? In fact sleep deprivation of more than 24 hours have shown people exhibit hallucinations and schizophrenia like symptoms. In short it can kill you, Rats totally deprived will die within 2-3 weeks. We spend nearly a third of our short lives in this state. My conclusion is we must do it to live. Back to dreams briefly! When we sleep and go into (REM) sleep our muscles become paralysed only to return when we resume to our awake or resting state. When asleep our bodies functions are slowed down, our breathing slows to deep long breaths.

Other than being in a coma it's probably the closes to being dead you get under healthy conditions. This is interesting at this stage because I think this is a point we get close to the spiritual conscious realm, we have let go a bit from our normal awake mode and physical connection to reconnect to our deeper selves. That is why NDE'S mainly happen in critical care in comas or serious lose of consciousness in this world realm! See it like magnets the closer you put them the more attraction and connection they have.

To dream is to be free
To awake here
I find the physical me.
PP

A matter that becomes clear
Ceases to concern us.
Friedrich Nietzsche

Nothing in life is to be feared
It is only to be understood
Marie Carie

CHAPTER 6

Evolving Consciousness

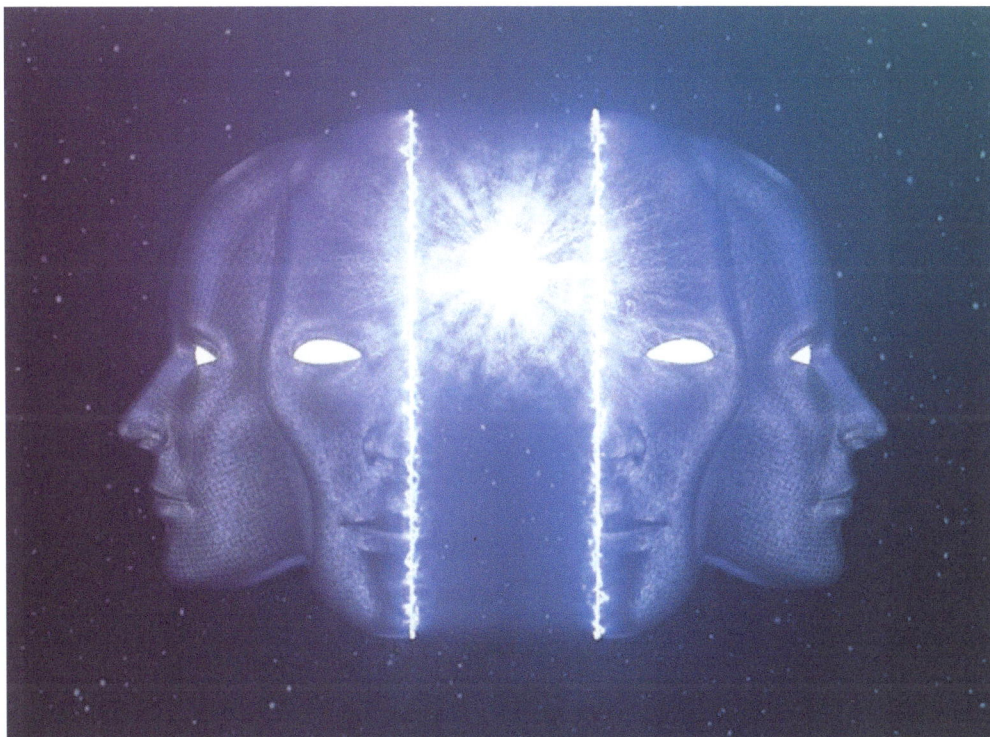

Consciousness in Mind

I'm conscious of my mind
The person you're find
The soul behind the mask
We are the soul cages
The spirit and the dream
My timeless soul
To evolve and grow
This eternal soul
Is forever to be
Collective to be one
A soul journey to make
Returning to the source
the consciousness of me
PP

Given the nature of the previous chapter it's time to delve into something wonderful, that is who we really are! In chapter 1 we talked about physiological and human behaviour, How everything is subjected to changing states of humanity. James Lovelock coined the Novasene the time in which we are evolving into. We will cover later with bio-engineering and cybernetics. These subjects are evolving and so too is our consciousness. Try thinking of the universal consciousness as you're own mind, think of when you were a baby, a small child, and then an adult. You're mind becomes more powerful as you grow, it expands allowing you to become more perceptive: more so with an open mind to clear a path to connect.

I believe that our universe is influenced by another realm. In that there is a cosmic other world not in a planetary sense! Where there is a very ancient intelligent collectiveness that may of Been somewhat like us once? In that realm "THEY"! Evolved to be super intelligent, going from rudimentary stages becoming environment changers to eventually being super collective with the ability to create universes! Here in this realm their influences play apart is this universe, I mean it's their physics.

We are off them, we are the eyes of someone else having human experiences. I believe this intelligence has transcended the physical to become pure energy consciousness or even evolved from a previous one? As I have said before we are just one of an infinite amount of them and only stands to reason they can be connected by a fine weaved interlaced exist's.

The universe is expanding and we know that there was a Big Bang because we can see the evidence of it in the cosmic background radiation left after it. Edwin Hubble 1889-1953 an American astronomer discovered that the faint fuzzy patches and clouds that were called nebula's Weren't all gas clouds but many were in fact Galaxies and giant stars clusters. Some of these galaxies would vary in brightness and were classed as (Cepheid Variables) the brightness is due to exploding stars or supernovas and we can get a good indication of distance. We can tell a lot from the light in the spectrum. Red shift is going away and blue is moving in our direction. In 1998 researchers discovered that by observing these supernovas, that can be as bright as millions of stars. They were fainter than they thought and thus traveled greater distances. This discovery lead to the conclusion that the expansion of the universe is actually speeding up, this was a surprise as it was thought that it would be the opposite? We have a good picture of the expanding universe starting from a single point. All matter, space and time came from it. In all this time it's Been expanding but into what? We have established there was a creation of some sort before the universe, for it is here! And something else we are expanding into Helping to support the multi-verse theory. By thinking this I feel it makes sense that some dimensions are closer to us than others. We are a way for the universe to know itself, before matter there was energy. This is crucial to understanding the universe and ultimately us.

We are not what we seem

Buddhists say the more you search for yourself the more you will not find it. When we refer to our body the self becomes the owner. When we say we are old this is normally referring to the body or I'm tired. Physical and emotional feelings come hand in hand, you can't just feel physically tired without feeling it in your mind I have now observed my mind by talking about it.

Cogito ergo sum

I think therefore I am, The self is aware of all things Material but when you look for the mind or ego which seems at first to be in command of you, you can't find it? It is just a programme of information processing. I read a description and work it out and then feel a relation as if there is a ghost in the machine monitoring me, like managing you're home. In the quiet time when you are at rest you're inner body house may look empty but all the house work has been done! The bed has been made our shoes have been polished and the breakfast has been cooked. All not seen but nevertheless it's done! People say seeing is believing however that doesn't mean it's not true. You can't see love directly but you know that you do! You sometimes cry but shed no tears. We live in an apparently physical world with ground, walls, water and the table but this is just an illusion of being solid.

Matter is really mainly or even just empty space? When you look closer at the atoms you find they are subatomic particles like a grain of sand moving inside St'Pauls cathedral. That is a considerable amount of space in comparison and the same goes for the smaller particles and so on! This conjures up an impression of nothing: however this is in fact false there is something we call it energy. I call it life or spirit! I spoke earlier about matter, walls and stuff! Take this table I'm writing on for instance it feels solid to touch I can bang it with my hand which I have just literally done! Please do the same to get an effect of what I mean. If you don't have a table try a wall or even the floor! Give it a few moments. . . . . Feels solid nothing usual there, it always does. This is the fascinating bit what you are actually banging is strong force gravity or as physics calls it, Basically energy holding it together. Imagine the wall or anything solid is a piece of chipboard held together with glue. Now imagine removing the glue like clicking you're fingers what do you think would happen? Yes, of course it would fall about! Back to my table or you're wall and in you're imaginary way click you're fingers again and remove the energy from them. Again they would lose their solidity, structure and form entering a state of entropy. They would lose form. We are all held together with this cosmic glue, We call this the laws of physics. The universe is governed by these laws! To me that implies pre-determinism some kind of cosmic plan.

I will cover more later on with transcending further in the book. Saying here again because I think it is fundamental and at the core of the book! We are spiritual evolving entities in a world of holographic material that seems solid and seems real Because we perceive it. What I'm proposing is we are from another realm and interlaced with this one, we are the progeny of an ancient cosmic precursor, in short we are an extension of them and in essence we are!

In psychology the notion of self is to recognise ones self as a persons experience, as a single autonomous being separate from others and all other objects. The self is conscious of one's physical body plus conscious of our minds! (me) my body, myself includes the mind and (I) the observer of them. When you say you're Leg hurts it is you're mind telling you or informing you of the location of the problem. The mind is the receiver and calculator of your physical being. This is the bit that gets harder to conceptualise the (I)! Here we go, I wonder about whats going on in my mind? I am talking as I'm separate from it. In the sense of being an observer to it. Having a close relationship with it appearing to have an outside view. My will is to do something my mind is trigged and then controls the senses to make the body move. To me I would say consciousness has another state but also has a close relationship with the mind and body.

CHAPTER 7

The Tears of Gaia

The Tears of Gaia

The tears of Gaia fall
They fall in torrents from a weeping sky
She pleads, she bleeds, she cries
The biosphere to the stratosphere
There's a crimson sky
theres a forest fire
Pleading asking why
Shall we make snow castles
In the polar white
But I fear it's already here
I see them in the rains of Gaia
I See them in the flooded plains
I see them on the marching sands
For they carve the changing lands
The blue line circular
Standing protective against the void
The coat between the cold of space
Shall we protect our only place
PP

First of all let's start this chapter with a little history about Gaia and the Gaia theory. The word Gaia comes from classical greek mythology of a Goddess and from the poetical form of land and earth. It can also be spelled (Gaea) It comes from the ancestral mother of all life. James Lovelock the father of the Gaia theory, was a planetary scientist, climatologist and inventor, Namely the inventor of the Microwave and the ozone detector. Like Charles Darwin a true pioneering thinker. He came up with a theory that the earth acts like one large organism, a self regulating and sustaining system putting life as the major contributor to the worlds atmosphere. At first this was not received well by the scientific community but after many years of science and research today most are now on-board. He is an amazing man still thinking and writing at the age of 100 years old! Publishing a recent book (Novasene) and at this age "WOW" We are coming out of the anthropocene were we humans used Steam and Nuclear power. The Novasene will bring forth the AI and Human collaboration or even supersede us, who knows? I think we will be apart of it. Ok the hard fact's for life on earth and our future! The Sun and earth are old and if for a moment you can imagine something wiping out all life we probably wouldn't have enough time to evolve again. Heat is the problem for our world. First let's start with our Sun, a main sequence star well into its prime. There are numerous small stars like ours with average ages going on for billions of years, however our sun is billions of years old.

Stars like Humans don't live forever and get old too, going through changes like us! When our Sun uses up all its hydrogen it will begin an expansion. The sun is hotter today then it was when life first appeared, the Sun is about 5 billion years old and in all that time has been converting hydrogen into helium but this will end and put a strain on our star! The Sun was about 70 percent as luminous as it is today and slowly getting hotter and brighter. As time goes on it will one day eventually become a red giant expanding past the orbit of the earth, by this time life on earth would have long gone extinct. Don't worry we have a long time to go and hopefully we would have colonised most of the Galaxy and become a truly cosmic intelligence. It is estimated that this red stage will take billions of years to be a problem but we have still got a more pressing problem on a smaller time scale that could warm the earth to be dangerous for us here on the Gaia ball called earth. In our relative short time out of nature only a little over 10 thousand years we have started to change our planet and climate.

The rise of Man

Homo sapiens prior to the last ice age which ended around 11.700 years ago and started some 2.6 million years back had a negligible effect on the earth! The last ice age that ended the Pleistocene saw Homo sapiens "US" evolve and start to make a mark and leave our impressions on the earth, becoming wide-spread across the globe. Unfortunately our ancient cushions The Neanderthals never made it through and the last known ones went extinct about 40 thousand years ago, their fossils were found in Gibraltar in caves. With the ice sheets gone it paved the way for the only surviving species that was us! Not having much of an effect on the land before this, it was all going change with domestication of crops and animals. We started to settle down from our hunter gatherer life styles. We saw the advantage of farming and breeding live stock, this produced a surplus of stock and was a pivotable turning point in our presence on earth and I guess this is where we can trace the start of the damage we would inflict on Gaia. Prior to this we travelled light making very few products, quite clearly because we would have had to carry them on our travels and fast sometimes, when moving to keep up with animal herds to hunt. Our populations were small while we were nomadic. It was hard to guarantee our survival from day today with what we could find and hunt.

Our time had come, No more were we at the mercy of nature and severe weather conditions. We built habitats to shelter from the storms and giving protection from predators, we penned animals giving them protection. Basic farming had arrived, we sowed seeds and grew crops. It is thought that the first crops were cereal type grasses that we could make bread with. They were cultivated by man in the middle east about 8000 BC. Life had become fruitful and we started to thrive! Having more offspring and higher survival rates.

We had come of age, the charting of the ascent of man became easier because we started leaving marks and you can use these to trace our progress. As we left our signatures with increasing impact on the environment we made settlements, then villages, towns, to cities. People became hundreds, thousands to millions and now with a world population of over 7.5 billion. The exponential growth goes hand in hand with the demand for nature resources. Take in for a minute the world is a finite place and is not an infinity sack of ever giving! While this revolution was taking place so was another in us, we started creating earth gods, sky Gods even animal Gods. We developed Art, writing, and maths. finding many ways to express ourselves we started to feel a larger connection to the world! It is not surprising we became spiritual. Not just wanting to survive we wanted to work it all out, and then came forth the birth of science, this science would revolutionise our capabilities moving towards and then into the Anthropocene. Now there has been a number of time periods proposed ranging from the beginning of the quaternary period some (2.6 million years ago) to the late 1960s but I'm going with the start of the agricultural, domestication and human settlement. Quite clearly every stage in the past 2.6 million years have had important pivotal points in our ascension out of the ecocentric place. Recent changes now include global warming, habitat loss, namely deforestation, chemical pollution in the atmosphere and alas the extinction of countless species.

The age of Steam

We have now arrived at the industrial revolution and the take over of mechanical power over man and horse power. Love him or loathe him? Thomas Newcomen (1663-1729) was to be in fair statement the father of the Industrial revolution. He was a baptist lay preacher, blacksmith, and engineer. He invented a machine for raising water by means of fire, it was the first pump. As there was a population increase and a high demand for raw materials our need for goal that was the wonder fuel of the time, it would be for hundreds of years! There was a problem coal had to be mined, goal was needed to smelt iron among other things to make building materials, and great steps in construction emerged. However the mines would flood and took time to manually pump out the water. Newcomen with his steam powered machine speedup the extraction of the water. Food production also benefited from steam power, for thousands of years our bovine friends were, if you pardon the pun! The main horse power. Horses were the main pulling power for ploughs but this was time consuming and tiring for these strong beasts. In comes the tractors, more powerful, runs for longer without getting tired and as long as you maintained it, it would just keep going. From this point forward the world would truly be in our hands.

The problem

Before we continue let's get the main cause of the problem out of the way in a few simple words: simply down to us! Unavoidable, unequivocally plain and simply we are the main cause. Our sheer numbers are clearly out of balance for our size and food requirements. A few number facts! There are estimates that range between 150-330 million of Human population globally on the onset of the agricultural and domestication period. The current world population is put at 7.6 billion, by 2030 estimates of 8.6 and by the end of the century 11.2 billion. As you can see there is an exponential growth. This should not take a rocket scientist to workout that this can't keep going on forever, it is just sheer maths and logic! Chimpanzees are our closet cousins by comparison and have an estimated number of around 170.000-300.000 and didn't take our path in evolution and have a negligible affect on the environment. We on the other hand have not helped them by our encroachment and the loss of their habitat! So whats going on? The Gaia theory in principal is a world of interaction with life, a kind of symbiosis between organic and inorganic. A good way to perceive it is to imagine the world as one large organism that acts as a self-regulating system. Gaia now days has an inclination towards a cooler climate. Life appeared 3.7 billion years ago: give or take a few sun rises! The atmosphere then was very different to ours today.

The earth formed 4.5 billion years ago and taking millions to settle down and cool to be stable for life to begin. The atmosphere of the time was mainly carbon dioxide and other noxious gases due to volcanic activity. There was vey little free oxygen and it was locked up in the ground. The first bacteria lived in a none oxygen environment and slowly transformed the earth by adding oxygen as a by product into the atmosphere. Water carrying comets bombarded us depositing this water to help make the oceans: It was time for life! Cyanobacteria were believed to be the first life on earth. Living in a very different world these photosynthetic organisms started to multiply and evolve. These single cell blue-green bacteria's would be the plants and too the evolution of the animal kingdom! These plant ancestors created the oxygen we have today and are still doing it! Taking in the carbon dioxide from the atmosphere and water from the ground and a dash of sunlight "hey presto" photosynthesis, in this process it would make the sugars and from the water release the oxygen and the sky we see today and the Air we breath.

Along comes man

The sky is changing again due to us, scientist have been keeping records of the weather and temperature for hundreds of years but we use records from 1880 onwards some 137 years because we had a reliable accurate global accounts.

Today we have a myriad amount of data on just how fast the earth is warming. I have seen the change in just my own life time! So what are we doing to make this happen? We are of great numbers and where it is possible to live Humans are there. Half the worlds habitable land is now occupied and used for agriculture having a massive effect on the environment, that is endangering bio-diversity. We to our shame have been responsible for so many extinctions, in fact it is said we are in the latest great extinction period. The last one was the KT 65 million years ago in the cretaceous when a giant asteroid that largely contributed to the death of the dinosaurs and countless other life forms. With this one you can see the culprit at work. If you want to see it take a look in the "mirror" and you're see it looking back at you! Here is an alarming figure, a new study has put only 5 percent of the earths landscape untouched. Here are some reasons.

. Agriculture
. The physical extent of Human settlement
. Transportation, railroads and highways
. Electrical infrastructure ie power lines
. Recreational sports and parks

Livestock

Just imagine if we doubled the demand to match the human population of say 14 billion, what kind of world would it be? Would there be any natural forest's or untouched beautiful landscapes?

Human City lights Viewed from space

What a remarkable image above of the lights created by man and if you look carefully just above the curve of the earth you can see the thin blue line that is our life protecting atmosphere. On looking at this photo a thought came to mind of just how it reminds me of a virus or a cancer spreading, Thats a sobering thought!

Deforestation can be easily seen from space too and comes as a direct need for land use and building materials. We have a taste for beef which takes up a lot of land space per head of cow. When you think of how much life would of Been there from trees, ground covering growth and fungus, there would of Been birds, insects, small rodents, arboreal creatures like monkeys, snakes and mammals like top predators, the list can go on! Very importantly species of plant that could have medical use to help cure deceases. Unbelievably more than 200.000 acres of rain-forest are burnt every day. The world lost 12 million hectares of rainforest in 2019. A Hectare is the equivalent to 10.000 square metres or 2.47 acres, let's put it in more laymen terms the equivalent to 30 football pitches every minute, and I just can't get my head around how staggering that amount of damage represents! Most of us live our lives in our immediate environments, seeing houses, roads, parks and places of work. Our world is generally seemingly unchanging but with what you have just read puts another perspective on the greater world to which we belong. Why so much hard clearance? A number of reasons but probably the worst is animal grazing on large scale Industrial farming! We Humans have a taste for meat especially Beef! Globally we consume around 350 millions tons of meat a year and is predicted to rise to 460 million by 2050. Burgers vs the trees! Who is winning?

It's not safe in the Water

70% of the world is covered in water 97% of that is salty and mainly in the oceans, Only 3% is fresh water, some 2.5% is locked up in the poles, glaciers, and in vapour in the atmosphere. Only 0.5% is available as drinking water. Another great Environment for food harvest is in the oceans and rivers.

Boy Have we exploited that environment! One of the most damaging fishing technique is dragnet fishing, in the process of harvesting the main catch you end up destroying loads of other creatures environments disturbing the sea floor disturbing corals and sponges. We have also depleted the fish stocks too dangerously low numbers. Apart from that we are also polluting the Oceans and Seas with raw sewage, oil spills, rubbish of all descriptions, and the infamous plastic waste! Pollution kills trillions of organisms in the food chain. We normally only think of whales and dolphins but when you think of the numbers dropping it is important to recognise the creatures they live on, the whole food chain as a whole. Rising sea Temperatures are also killing countless microorganisms. There is a mass extinctions process taking place on land and in our Seas as I write. Species are going out before we have had time to even document and record them.

Rubbish

We Humans produce copious amounts of waste, the average household waste thrown away is more than a ton and put together about 31 million tons in the UK Every year! The most common place our waste goes too, is land fill sites. Some does go to recycling centres, glass, aluminium, and even plastic. Unfortunately not anywhere as much as is needed! As recycling processing can be expensive. Some unscrupulous people and companies discard waste that is hazards to the environment and kills loads of wildlife. Just one chemical leak or something poured down a drain. Can take years for a polluted river to recover. Plastic particles are getting into the food chain, with the wide spread and use of plastics which we use in almost everything in life now days. These polymers break down and get taken in by the wildlife and has been linked to fertility problems and defects. In men or males also linked to low sperm count. Might not be a bad thing for Human population control!

The sky really does have a limit

Since the industrial revolution we have been burning fossil fuels like coal, charcoal, oil and petroleum. As industry and the need for home electricity multiplied so did the need for energy. The problem with fossil fuels or carbon fuels is they are very polluting. The use of planes, cars, power stations even home open fires puts millions of tons of carbon dioxide into the atmosphere. Carbon dioxide is higher now than it has been in the last 800 thousand years. We have been releasing so much of it in a relatively short period of time, As well as other gases in the last 2-3 hundred years that took millions to be locked up in the earth. It is a bit like opening a flood gate all at once. Methane from farming of millions of cows and sheep, co2 from cars and CfC's chlorofluorocarbons and nitrous oxide we have used in aerosol cans. Just think of when you turn the key of you're car or flick a light switch on, you are adding a bit more to the problem!

There is a long list of chemicals that can added to the pollution problem that has been pre-mentioned, However you maybe surprised to read that water vapour is a high contributor to global warming, it accounts for 60 percent of the earths temperature. The addition of none condensable gases we release into the atmosphere has lead to more water Vapour, making it a potent greenhouse gas. The more it builds up the more trapped heat stays in our atmosphere. The same way a greenhouse works when sunlight enters through the glass and then gets trapped inside to hold the heat. The planet Venus is a good example of a run-away greenhouse planet making it even hotter than Mercury, even though it is millions of miles nearer to the sun. We have a frame of reference in our own solar system to what could happen in theory and in reality. Seeing the increasing melting of ice at the poles and the breaking away of the ice sheets, it has alarmed the scientists just how fast it is happening and faster than they were actually predicting.

The release of viruses

The thawing of the permafrosts of the great tundras in places like Siberia, and great mountain glaciers are also happening very fast releasing more carbon and the other gases, also pre-mentioned water vapour still adding to global warming. Just a breather for a moment .... Ok' with longer warmer summers and shorter warmer winters this water is not returning to its Frozen State, some melting glaciers in places are almost gone in recorded history and we are watching them go before our very eyes.

There are diseases hidden in the ice

In light of the covid 19 or coronavirus a new disease, a global pandemic we are being faced with, something new or old? For thousands of years bacteria and viruses have been frozen in ice and permafrost and now as the world is warming they are being exposed to the modern world and becoming reanimated. This warming is melting the permafrost releasing ancient viruses frozen for thousands of years that we have not been exposed too and have no natural immunity off because we have not recently had to live with it. In august 2016 in Siberia a 12 year old boy died and 20 people were hospitalised after being infected by anthrax. It is thought that some 75 years ago a reindeer infected and frozen with anthrax died! It became trapped in frozen soil as permafrost. In 2016 we had a very hot summer and it thawed out. This corpse released the anthrax into a near by water source and then into food supplies infecting Two thousand reindeer also infecting Humans. The Spanish flue back in 1918 infected 500 million people and is thought to be a historic event and now gone, but it still exists in frozen form and we have them frozen in labs! As more ancient soil thaws out more diseases can be released and we may see more covid like pandemics.

Water finds it's own level

So where does all this water go? It goes into the atmosphere and what goes up must come down! More rain means higher river levels and subsequently more flooding. Combine this with sea level rise many low lying lands, marshes, estuaries, Paddy fields and fertile farmlands could be lost to the sea! Millions of Human habitations are built near rivers and low land areas and we stand to be displaced as water rises. This displacement of millions of people by the end of the century is a real and pressing issue. Some scientists predict if we continue at this rate we could even have an ice free Arctic by 2040? The Greenland ice sheet is a marker for predicting sea level. If it melts we could see a twenty foot rise is the seas. What will we see if this happens? First we will feel hotter and many plants and animals will simply disappear! The world will have increasing decertification, I mean just look at death valley eastern California, it has regular temperatures in the upper 40s degrees and is almost devoid of life: Hence the name death valley! This is not unique to the land, it happens in the seas too (Aquatic deserts). I think this may come as a bit of a surprise for most! As we are land creatures we don't notice this almost alien environment. When we go on holiday destinations one of the main attractions that people look for is clear blue seas. Inadvertently choosing a sea of limited life, Yes you do see sharks and rays and around reefs and pretty fish but for square metre area of waters nothing like as much as abundant as the northern seas and oceans. Remember large creatures feed on smaller small ones, It is plankton and other minute animals and plants in the water. That is why the seas look grey or darker blue because they are more heavily populated. You wont find whales in the dead sea but you will in the cooler waters. Life likes it cool and here in-lies our problem because we are making it the opposite! The warmth is killing the aquatic life. The next chapter will see how we can address these problems. A bit of soul searching is needed and a look to our future. If we do really want one? We must take note of the clear evidence and act upon it.

"The future of Humanity
And indeed all life on earth
Depends on us"

Sir David Attenborough

CHAPTER 8

The Gaians Unite

To start with I would like to say I hope the previous chapter has put our position on earth clearly as critical and urgent, but not irreversible. Gaianism is basically a kind of religious philosophy putting Gaia or mother earth at the centre and not us! It sees the earth as a kind of super-organism, but not a sentient being. I use the term Gaia because I like the analogy of a kind of mother and for all intents and purposes may well be! That is as far as I subscribe to the philosophy. The mechanics of how the earth system works, and the great work of James Lovelock in his Gaia theory holds true! As for the mother only concept, well that is only partly correct with a different perception of it. This is Due to our static position and dependence on the earth. We are spiritual beings evolved out of nature and at least for now we have a close connection to our earth, However this is not our ultimate future, we spiritual beings are on a conscious journey through time and space. Try to see our world as a nest, we get to a point and then we must fly! An analogy could be the earth is our old mother and we are young; it's time to look after her. In this last statement I say young and yes for here we are indeed, we have only really just begun our great journey.

First population Control, probably the most problematic and sensitive subject in this chapter, and I hope I don't offend anyone in this section! But it is crucial and an unavoidable issue, Quite frankly its our largest problem. Most of all I have written this is the most difficult topic I have had to address, I have been racking my brain to come up with the solutions and this is what I have concluded, but always open to new ideas?

People management

As a world community we must agree on a top level number per Square Acre of land and put parameters of population control in place. In simple terms numbers that can be supported without harming ourselves or the balance of the world environment. Birth control, incentives not to have children, this could be financial. Draconian as it may seem but some kind of measures for offspring limitation laws will have to at this rate sooner than later be implemented. Don't you think having six kids in this day and age is somewhat excessive? There has been systems and methods in nature long before we came across the problem or even long before we were here! If an animal species number started to come out of balance in their environment there would be more predators taking advantage of the ready supply of bounty on tap. Starvation, a higher birth mortality rate and some would migrate to new pastures. They didn't have red cross feeding stations and air drops to support them! The plain fact is some would survive, and some would not. Nature balanced them out in one way or another; even extinction! Does this ring any alarm bells? It "should" We're not meant to live in such high numbers in deserts. Only with extreme measures and external help is it possible to live in such environments.

All we are doing is escalating the problem and sending the wrong message. Maybe it's time to start stepping back and let the true sustainable numbers return to the regions? I know it's easy for someone in a more fortunate position to say that, but it doesn't detract from the fact that it is true! Going against what we see as Human caring and our ideas of human rights. Human rights won't save our world and ourselves! Human common sense and people management, might just do it! This line of thought I take from logic of the needs of the few, as in sustainable numbers outweigh's the needs of the many. If we implement these measures with a backup plan, which I will cover later in space colonisation in Spearhead to the stars and Cybermen, we have a chance and a great future ahead in this universe. We do however have to get onboard and make the changes and the inevitable hard decisions like the forementioned. We must untie our hands of old habits.

A clean future

We have covered the carbon fuels in the previous chapter and now we can look at alternatives, It's good news! All of it is mainly on tap and we already know how to do it. Already we are starting to use this technology and it's much cleaner and better for the earth, thus better for us! Not all Human developments are bad and along the way great minds have dreamt of endless and cheap supplies of energy, and how to mitigate the damage.

The Sun

The sun is a powerhouse of free energy and we as the recipients of this use it to be converted into calories from our food. This is a measurement of energy in biology, and is stored in seeds, then plants Transferring this to animals. We really are the children of the sun! The sun releases energy at a mass conversion rate of 4.26 millions tons a second, this is the equivalent to 384.6 septillion watts of energy. I had to do some research to what this was in old English so I could try to relate to it, it translates to a 1 trillion is 1 with 12 zeros and a septillion is a 1 with 24 zeros after it. Now i'm a bit lost now myself!

This is every second would you believe? An astronomical number of power radiating out. When you feel that warmth on you're face, how often do you think of that power? With all this solar energy on tap it only makes sense to use it as we are increasingly doing so, but we have to speedup! Solar panels on roofs are saving people money and even selling the surplus electricity back to the national grid. I think it should be mandatory to build new homes with solar panels as standard. So should the street lighting, road signs, meters, and all manner of vending machines. Another good idea I have seen when on holiday is water tanks on roofs to warm up hot water from direct heat from the sun, it's absolutely free. There must be a myriad of solar possibilities we could power devices? Massive greenhouses on farms and air-conditioning which we will

Free power from nature in our hands

Need more of as the world gets warmer. The sun has always given us life and so to can it help preserve it! We also can reflect unwanted heat by painting white on roofs, sides of buildings or using reflective foil in built up areas, where there isn't much greenery, as plants absorb a lot of sunlight. Policies to be put in place for keeping our cities and urban settlements organic and green as possible. I work for a company "Urban planters" which is part of Tropical innovation Ltd, we construct living walls, green screens and all manner of planting formats to roof top gardens. We believe in Biophilia which is an innate affinity with life and living systems, the term was first used by Enrich Fromm, who described a psychological connection and attraction to nature. The colour green is the most calming colour in nature, not surprising for we came from it not too long ago! To me it's a no brainer, keep what we like and keep what we "need"!

There is even a plan of using a low earth orbital screen block, satellites to block a percentage of light and heat away to help cool the earth or heavily populated human cities. When you are sitting in you're garden and a cloud or even and aeroplane passes over you get a shadow this is what we can do, it's basically eclipsing the sun.

Wind power

This is free too! "Yea" Humans have been using wind power for thousands of years sailing around the world or windmills for making flour, it's an old Technology. So what took us so long to see its full potential, and only just recently start to see it as a viable energy source? Well, I think more than one reason but things moved slower in the past and then we discovered the great wonder fuels coal and oil. The wind turbines have come of age and fossil fuels have had their day! Unfortunately there is no getting away from the fact these huge carbuncles or blots on the landscape really are not pretty. However we can put them out to Sea or in places on land that are already not scenic.

The very first wind turbines to produce electricity was created by professor James Blyth 1839-1906 in the year 1887. It was 10 metre's high and had a sail of cloth, today they are becoming very efficient and a lot better than they use to be only a few years ago! What I would like to see is schools, office blocks, industrial sites and even shopping centres having them as standard. So you may ask what happens if there is no wind? Don't worry power can be diverted from other area's and other means of power production.

The Tide is turning

You can thank the Moon the next time you look up for in part creating the tides! Just think how grateful surfers are that waves exist? The Moon orbits the earth in a Lunar month 29 days 12 hours 44 minutes and a few coughs LOL! And this exerts a pull on the earth by means of gravity. This causes tidal forces a bit like a dragging effect. A good way to see this is how a smaller object can have an effect on a larger one is to watch a Hammer shot athlete spin around. If you Haven't noticed before? Take a closer look and watch how the person wobbles as they turn! This is because there is a transferable force of gravity, little as it maybe it still exerts an effect on the athlete. The power of nature again free for us to use! If you have an automatic watch it uses a method of charging up the spring by a rotor or weight That oscillates freely to wind the spring, the power house is the human movement. In wave power the same can be done to produce electricity, the power house this time is the moon and wind transferring the wave power into energy. Just think how many coastlines there are in the world, when you walk along the beach just how many of us can see the great power at work as we stroll? There is another power source behind waves and that is from the wind that is generated from solar energy and guess what? It's all free again! A average 4 foot 10 second wave striking a coast can put out more than 35.000 horsepower per mile of coast. We call this hydro power and has been used in rivers to turn watermills and great dams that have been built to power the grids. The largest hydro dam built is in the UK is dinorwig power station in Wales, It has a total capacity of 1.728 megawatts, which is enough to power up to 2.5 million homes, "WOW" combine the pulling effect of the moon one side and the sun on the other, causing a bulging process which intern makes tidal waves and the human technology of using water, Hey presto! an almost endless supply of power.

It is pleasurable, when winds
Disturb the waves of a great Sea,
To gaze out from the land upon
The great trials of another

Lucretius Pompelii 94 BC

Nuclear

This power source has a lot of stigma attached to it and Generally has more of a fear factor. We are back to atoms again, for this is where we get our power from in Nuclear power stations and unfortunately weapons. How does the nuclear process work? It works in the sun as (Fusion) May I add we have not yet managed to master that method! In the Sun however the process of fusion produces heat. Under great gravitational pressure atoms fuse together, Hydrogen transfers into helium, In our star the rising temperatures start the reaction. If we manage to tap this natural wonder of almost endless power the future would really be our oyster! The system we are using today is nuclear fission and is not so energy efficient as would be fusion, but nevertheless still a great power technology. It works in the opposite way by splitting the atoms, More of a runaway method and not as controlled as fusion, just like petrol there is waste. What basically happens is we use a fuel like plutonium or uranium and we split them to release tremendous amounts of energy to heat water into steam and turn the turbines to make electricity.

This atomic power is a chain reaction inside the reactors adding to the world of ever needed power, the good thing is it will last a lot longer than coal and oil supplies. As I grew up I was always aware of the public perception of nuclear power and the accidents that have happened, such as long island, Chernobyl and the Fukushima 2011 nuclear disaster. If care is taken and no more major disasters, its a much cleaner power source and compared to other fuel industrial mining types is the safest. Here is a list from 2012 a year after Fukushima of mortality by power source.

. Coal 100,000 deaths
. Natural Gas 4,000 deaths
. Hydro 1.400 deaths
. Wind 150 deaths
. Nuclear 90 deaths

This list is based on mining extraction and the use of the energy. The figures change from year to year, but Nuclear power is still considered the safest fuel source for power. Compare this with "Smoking" that kills more than 8 million people each year, 1.2 million of them are none smokers or passive exposed to second hand smoke, I guess you could include open fires? Based on what I have written I think the so called demon power is still our best power for many years to come. The time has come for us Gaians to unite to save our world, our only space platform.

The faster we stop fighting each other the better, we have a common war to fight to save our world. International cooperation is needed. We don't have invisible walls at borders between countries! The atmosphere doesn't care what we call England, the United States or the Brazilian rainforest! The problem is global and the fight is also global. It might not seem very spiritual so far in this chapter: but it "should" We are a privileged species with so much consciousness and a shed load of intelligence, we have been taking so long to see the threat to our future. You may say or think that the problem is too epic for me to do anything about it. A termite doesn't say I'm just one to make a difference or a grain of sand to make a beach. We should think; I in a collective can do my part! First we have to recognise the problem in us and others. This is what I'm going to call the (ditto recognition) the first step. The second is (unification cooperation) and the third is (ecopsychology) this should be on the school curriculum. We all should have a mantra: what can I do not to harm the environment in my daily life. If there is to be one religion to believe in, it should be (environmentalism). The world is losing habitats at an alarming rate and we are now in a man made extinction unfolding that could equal the one 65 million years ago that wiped out the dinosaurs and some 75 percent of life! The next part I shall list the extinctions of the past, and show what was not in the control of an intelligent sentient being such as ourselves. There have been five major events in the past.

. First the Ordovician 440 million years ago, 85% of Life went extinct.
. Second the Devonian 375 million years ago, 80% of life went extinct.
. Third the Permian 250 million years ago, 95% of Life went extinct.
. Fourth the Triassic 200 million years ago, 50% + of life went extinct.
. Fifth the Cretaceous 65 million years ago, 75% of Life went extinct.

The previous list is a stark reminder that change can be fast and catastrophic and out of life's control, from asteroids, volcanic eruptions and rapid climate change like ice ages.

The sixth mass extinction

Cause: Man

It is believed that we are now in a major mass extinction and we are the suspected culprit. A number of species have disappeared since our explosion on earth, and we see a direct correlation between our settling, hunting and the subsequent disappearance of species. All the previous events were natural and life had no control or foresight and was subjected to external influence, taking whatever had been thrown at it. This time it is different! Not only knowing what it is, we know it is us, thus what to do.

This time life can make a stand, we Humans can search our souls and think what we are doing, thus we can do something about it. The first time in the earths history has there been a creature capable of doing something about it, capable of caring for other creatures and perceiving a long term future for ourselves. This time the Gaians can fight back.

The Gaians fight Back

We can turn the tide of destruction, it's all about the changes and of course some sacrifices. We need to start large scale tree planting to replenish our forests. You would be amazed just how much life depends on them! An oak tree can support over 280 different species of insects, mammals, and birds in one species of tree alone. What else can we do in respect to trees? Here is a good idea? Just think of all those flowering trees planted just for the aesthetics, shrubs too just grown for our pleasure. We could start a planting program of berrying and fruit plants that would be more beneficial to wildlife and us! We would plant them along highways, parks, railways or even the street you live. Mindfulness to what we throw away. Can there be another use for it, does someone else have a use for it? As a grounds maintenance supervisor on a 23 acre commercial site, with woodland, wild meadow, pond, and formal planting, we have an environment ethos and love to be as wildlife friendly as possible, recycle what we can.

We won the Surrey wildlife Gold Award two years consecutively in 2018 and 2019. We did this with some initiatives like making use of materials on site that may have been thrown away And a change in the maintenance was part of it. Here are a few things we did!

- Allowing some formal grass areas to go to wild meadow.
- The use of wood pallets to make bug hotels, cables reels to make mammal boxes and old fruit boxes to make Bee and insect habitats.
- Coffee granules get bagged up for staff and also composted on site.
- In places around the grounds we have recycled plastic tables and benches.
- Inside the building we have recycle battery and crisp bins.
- We have done away with one use cups and polystyrene lunch boxes.

Most of these examples can be used at home too; if not on a smaller scale! But if everyone did it we would make quite a difference on holding back the environmental damage. By not doing this we would be adding more to transporting and processing which adds to our carbon footprint. We can also make our products last longer because every time we get something new we have to get rid of the old. A lot of manufacturers don't make products to last long because it is not good for business, However its not good for the planet either! I think we need to limit were possible the built in obsolescence. A prime example is my old computer I stopped getting updates but other than that it works fine?

So we bought a new one, however I didn't throw it away I use it for writing and storing photos.

Food on the table

With the rise of population comes the rise in food demand and as I have mentioned before there has been an alarming amount of land cleared for cattle, which is adding methane to the atmosphere. It is 28 times more powerful than carbon dioxide as a potent greenhouse gas. Grazing this cattle takes a lot more land and resources in comparison to crops. Answer to the problem eating more or a total vegetarian diet would help reduce the effect on global warming! When we farm large areas of mono-crops and graze beef we are denying the land of Bio-diversity in part contributing to the mass extinction that is unfolding before our eyes! Do we really want to live in a world of just cows, chickens and pigs? It's time to stop so much waste going to land fill sites and increase the building of recycling plants. The UK produces more than 100 million tons of waste a year, in 8 months it would fill Lake Windermere the largest and deepest lake in England. It is estimated that up-to 80% of waste we throw into our dustbins can be recycled or composted. I must say I have a bit of a bug-bear about "junk mail" being fly tipped through my letter box, even through I have a clear sign saying no "junk-mail" please! But through it goes (growl) I mean what part of that don't they understand? 17.5 billion pieces of junk mail are produced in the UK alone, and most of it goes to land fill sites. It's easy to stop, just bring in a law to stop advertising letters and leaflets and classifying it as fly tipping. To add insult to injury you are paying to get rid of it through you're council Tax for something you didn't want! I think I have made my point.

Just a few ideas

. Implement an Equilibrium Policy Directive (EPD).
. Every new build should be energy efficient.
. Slow down take away food with its waste.
. Look for and use bio-friendly products where possible.
. Reduce travel as much as possible.
. Make things last and where possible more than one use.

There must be lots examples and I give you a challenge to find and implement at least one new environmentally initiative, for now I shall leave it at that. Putting aside all this data and facts, there is something very profound about our predicament: Quite literally our survival! I believe we have a great future, we just have to survive long enough to live on other worlds. For now this is our only platform and at the moment the only place for life we know. If we are the only sentient beings and our world is the only world with life in the universe or just the galaxy then we are very rare indeed. We are the universe trying to understand itself, trying to survive itself and to expand and evolve itself! Wouldn't it be tragic if we went extinct before finding the answers, finding who we could be? We are the products of 13.8 billions years of universal evolution, and out of all the life here we have been given the gift of awareness, spiritual consciousness. So my fellow Gaians lets unite for the true fight and fight for our survival. I believe the answer to this material problem lies in our deep consciousness, only when we realise our power and energy to make the change in our spirits, in our souls the realisation we our much bigger than this and do what we always was meant to do!

CHAPTER 9

Spearhead to the Stars

Many years ago
The British explorer George Mallory,
Who was to die on Mount Everest,
Was asked why did he want to climb it.
He said "because it is there" Well,
Space is there, and we're
Going to climb it,
And the Moon and the planets are there,
And new hopes for knowledge
And peace are there.

- John F. Kennedy -

From the dawn of time man has looked up and wondered, what is it all? What would it be like to go there? It has been a topic of romance, poetry and many tales of adventure. H.G.Wells 1866-1946 novel The first men in the moon tells a tale of an inventor who wants to go to the Moon, Written in 1900. Humans have always been fascinated by the starry heavens and Gods in the sky! It was probably not until Galileo Galilei 1564-1642 who was an Italian astronomer, physicist and engineer used the telescope to show some of the wandering stars that we call (planets) were in fact other worlds. They too themselves had smaller worlds orbiting them, other Moons! This to me was a pivotal point of our view of the universe and it would never be the same again, it was full of possibilities. We had a new sense of our place in everything and places we could now go. With this in mind we were destined to leave our world. After great leaps in rocket technology in the 1940s We put Sputnik into low earth orbit on the 4 October 1959 and then Mans dream came true with the first Human to go into space: Yuri Gagarin a soviet airforce pilot onboard Vostok 1 on the 12 April 1961 fulfilled the dream. Never again would we be tired to the earth and we saw a clear path to the stars. H.G.Wells would have loved to of seen the next great step. On 20 July 1969 Neil Armstrong and Buzz Aldrin onboard Apollo 11 landed on the Moon. One massive step for an Ape descendant to step onto another world. We have now sent probes all over the solar system, landing on Mars, Venus and even landing on a comet!

We now have a permanent presence in space on the International space station, there are children being born on our world were our species has always been there! Having plans to return to the Moon, to put an orbiting space station in orbit around it and to go to Mars. We have always been wanderers and still we are inclined to go somewhere because it is there! You could say we were pre-destined to leave our world. I remember as a child at school I would stare out the window of my dormitory. I went to a boarding school in Surrey, England. I would lie there staring up as my bed was by the window and make patterns out of the stars. I would wonder what was out there! I reckon that was where my interest in astronomy came from? Latent inside me waiting to be released. Another thing out of interest, my dormitory was called "leith" and the names of the dorms in the school were named after Hills in Surrey: Box hill, juniper hill and mine Leith hill, etc. After many years and a bit of moving about I came to live near Dorking among the surrey hills, on Leith Road near Leith hill and my Doctors are the Leith hill practice, you couldn't write it, but I just did! Another one of those coincidence events that keep happening. One day in English class we were assigned a hobby project of our choosing and on our bookcase was a load of new books. We could pick one for our project. There was a whole manner of assortment of topics's, like second world war planes and tanks, football/sports, history you name it! Everyone grabbed the one that caught their eye and I was drawn to one on the planets and stars by Patrick Moore 1923-2012 the famous Sky at Night astronomer and presenter. On flicking through the pages I came across star patterns called constellations to my surprise. I used to try making shapes out of the stars myself! I read the book with attentive interest. The knowledge and imagery of the planets was limited to artist impressions due to that time, but that was going to change! With the space probes voyager 1 and 2 that were sent to the outermost planets. They were a huge success and discovered many new Moons and Rings. I like to think the spirit of Man was at least metaphorically traveling with those wandering space probes, and who knows one day far away some Alien race might encounter them and wonder who we are? Or we may one day travel so fast that we might catch them up and reunite with them and even put them in a museum? I am sure we will be putting flags on other worlds saying (we were here)! Even if we don't stay? The solar system is within our sights and with new technology and with on going developments we are working on, by the end of the century Humans will be living and working all over the solar system. We will be mining the asteroids and planets and their Moons. We're also learning how to survive on other worlds, learning how to terraform them in preparation for more distance worlds. Amplifying our chances of survival, we won't have to be so reliant on earth and will gain our cosmic independence. When we get to interstellar this is where we hit a larger problem! The solar system is big! But it's nothing compared to the distances of the stars.

Space station artists impression Home and work

The nearest star other than the Sun is Proxima Centauri, it is 4.2 light years away. Light travels at 186.000 miles per second, the sun is about 93 million miles from us and light takes 8.3 minutes to reach us, but from Proxima Centauri it takes 4.2 years. We are nowhere near able to travel that fast. Also it is said if a spacecraft did get to that speed it would become infinite in mass! For us to travel to deep space locations we will have to develop the means to do so with future technologies. First of all we have good news about "other worlds" besides our solar system we know of extra-solar planets. To date we have discovered more than 4.000 of them, and with thousands of possible candidates waiting to be confirmed? As our telescopes get better it is believed we will find a possible number of earth like planets, In size, with water and in the habitable zone. Sometimes called the Goldilocks zone. Hopefully with water in liquid form? There are about 200 billion stars in our galaxy, with 10 percent of sun like stars. Some 20 billion suns and if just a quarter of them have earth size planets with water, that could be a staggering 5 billion in our galaxy alone? This is good news for us with so many places that might support life as we know it, or with some adapting we could manage to live there! The downside of this is the massive distances, even to the nearest stars and there planets. The further out you go the longer it will take to get to there.

Terraforming

For the first part of Human colonisation of space we will start at home in our own solar system with planets like Mars, not too far away and a solid terrestrial world. Mars has a thin atmosphere unbreathable made mainly of carbon dioxide. We would have to release oxygen there and warm up the planet! On average the temperatures are minus 80-60 degrees a lot lower than earth. One bit of good news is we have discovered water locked in ice. We would have to release more CO_2 into the atmosphere to warm up the climate to melt the water, by thickening the atmosphere we would then start to produce oxygen from the gases on the planet. That is the basics with a little more chemistry that's the foundation.

Although ambitious, in time we will be able to do it. Developing new technologies with more hands on experience, and a presence on a near by planet, Thus using it as a learning curve we will take this knowledge to the stars. Back to the problem of distance, We need to overcome the speed problem, even for local travel. The space shuttle could fly at a specific impulse of 242 seconds or 17.500 miles per hour.

Rocket Engines Here are the types of engines we may use. As we go through the faster a spacecraft could travel! Near Space travel for example to a neighbouring planet would be just like jumping from horse-drawn power to an intercity train from London to Birmingham. The further out in light years the more of a challenge it becomes, but in theory could be done. The main problem, believe it or not is us? Biologically we don't live very long and we are Biophilic creatures and space travel really is alien to us. Below is a list of engine power.

. Rocket engine   242-specific impulse
. Nuclear fission.   800 - 1,000 impulse
. Ion engine   5,000 impulse
. Plasma engine   1,000 - 30,000 impulse
. Nuclear fusions   2,500 - 200,000 impulse
. Antimatter rocket   1 million - 10 million impulse

The more we go down that list of technologies the better the chances of colonising space, and later on in the book we will look at how we can overcome the difficulty, even at these speeds with the length of time ahead we face an epic challenge.

The speed of Light

. From the Moon   1.3 seconds
. From the Sun   8.3 seconds
. From the nearest star   4.2 light years
. From the nearest galaxy 25,000 light years
. Across the milky way   100,000 light years
. From the Andromeda galaxy 2.5 million light years

Comparing the speeds of the rockets to the sheer Astronomical distances involved we have a long way to go. The great thing about a long time, presuming we survive long enough, we have a lot of it to achieve our goals. If we sent men to Mars today it would take 9 months to reach it! And of course longer as you travel out. We Humans live at best a little over 100 years but mostly under. This is problematic when you're looking at dozens or hundreds of light years. So how can we achieve this great feat? I can think of three ways, all will need to be solved in they're different and unique ways! If we want to be a cosmic race we must climb the ladder making break throughs with every step!

We must live a lot longer, I mean a lot longer! More details in the next chapter with Trans-humanism and bio enhancement.

We could make self-sufficient generation ships capable of holding many hundreds or thousands of people with their descendants reaching the target destination, meaning people may be born and die on the ship throughout the journey, "Not great" This is the one I think more likely and definitely more appealing, This is Cryogenics!

Based on us staying as Humans with part genetic or cybernetic augmentation or body replacements. We will look at this in the next chapter. Lastly to achieve the bridging of the stars we will have to go through stages of power evolution. The small list below was first conceived by Nikolai Kardashev 1933-2019 a Russian astrophysicist in 1964 came up with 3 types of civilisations based on power consumption, Here is the scale!

. Type one, a civilisation that uses all the energy from the sun that falls on the earth.
. Type two, a civilisation utilises all the energy the sun produces.
· Type three, a civilisation utilises all the energy of the galaxy, This is quite some prediction, but we are not talking overnight! And it will probably take many hundreds to thousands of years to do. We Humans are used to long time periods. For millions of years we were using sticks and stones as tools like chimpanzees do today. It took thousands of years to domesticate, build cities, develop nuclear power and develop The physics to eventually build the rockets. Today we seem to be evolving technologically faster, If this trend continues we may achieve these goals a lot faster. The question is can we as Humans? As I have said before we are in a cultural Lag as a species. We are one step behind our technologies. Let's take it as a given we save our world long enough! We can then undertake the great migration to the stars. First we must master (cryogenics) and be able to sleep these great journeys. By doing this we leave all what we used to be behind, It's a one-way ticket. Truly a new start and a birth of a new kind of human being. I am sure one day we will do this, and here are some ways we may do it! As for-mentioned we could go into deep freeze and be kept in Hibernation State to be awoken on arrive or in cases of an emergency. An animal was brought back to life from a 30 year deep freeze, Japans national institute of polar research resuscitated two creatures called "tardigrades" sometimes known as water bears from frozen moss 30.5 years old. The researchers collected them in 1983 from an Antartica moss sample. They stored them at minus 20 degrees C and then in 2014 defrosted them. Taking 29 days to return to what researchers called normal condition, the other one died after 20 days. These creatures

are classed as extremophiles and are giving us new hope for life on other worlds. Life is tougher than we first thought! Some animals have a natural anti-freeze like some fish and the famous wood-frog, That can survive subarctic temperatures. The key to this is preventing ice forming inside the cells and thus not destroying them, if we can do this than we could in theory freeze and revive humans in the future. In 2005 scientists revived 32 thousand year old microbes frozen in a pond, this out of interest was the time when wooly mammoths were roaming the tundras and we humans were hunter gathers. Even more amazing scientists managed to revive an 8 million year old bacteria lying dormant in ice again in Antarctica. Seeing it done as far back as 8 million years and compare the sheer time need to travel to the stars, this could be hugely advantageous for us. Ok let's say in another scenario that we don't find a planet perfect for us, say a complete identical world in size, amount of water and atmosphere quite close with the right chemistry we could use? Then one option is we could send "AI" probes ahead of us like a vanguard to soften the planet setting to work with small nano like machines to self replicate and start building climate changing devices. While the terraforming is taking place, the infrastructure for landing, water resources, habitation units and natural resources can be put in place. Say these planets are no further than 1,000 light years we one day maybe be able to send frozen embryos and DNA that could be cloned by Ai robots. One advantage of this is the new Humans wouldn't be home sick! They would only know one world. Multi generation ships could take countless Human lifetimes, unless we master cryogenics to achieve our goal. I would like to think the AI's would be as human like as possible because they would have to be mother and father!

They will have to be the teachers, doctors and even to start with the law enforcement, at least until we are firmly established as a fully functional civilisation. While on this journey and on arrival there would be continuous updates from earth probably daily, with news and technology developments transmitted to the AI,s and of course transmissions of progress of the colony back to earth. You never know one day they might send themselves onto new worlds or even back to earth? As I said earlier about a new kind of Human being, this is because genetically we would be engineered to live on another world. This world could have a different strength of gravity, the atmosphere maybe thicker or thinner, the distance from the parent star maybe different from the earths. The rotation and orbit time and the axis angle. This would mean the day and nights would be unique to the planet. In time we could be alien even to ourselves! Ok we can see there will be quite a bit of physical change but what about spiritually? From the first days we evolved out of the rift valley in east Africa when we started losing the forests our previous habitat, we had to embrace change. Human history is all about adapting and exploring. We as Humans have re-evaluated ourselves continuously and generally changed to our advantage. In plain and simple terms it's who we are! Our consciousness will continue on this journey, because it too will change with new experiences and understandings. Embracing our expanding world we will think Humanity into the stars.

Warp-drives and Wormholes

Let's go further, let's say the whole galaxy or even anywhere in the universe? How about outside it? As we have seen the further you go the longer it takes, and the universe is a vey big place. Even at the speed of light, as I have said the laws of physics stop an object of mass from achieving the speed of light. We may come close? Albert Einstein 1879-1955 a theoretical physicist predicted time slows down the faster you approach the speed of light. When you get to it time stops! This is what I am going to call the "Now time" Matter is energy vibrating slow, and high energy is vibrating faster, light speed is high speed and so too is universal consciousness, this is where time acts different as I have said. This is where true consciousness exists. Time has a limit for this universe but not in another realm, it's laws operate differently to ours. There maybe a way around this? If we can't beat the physical limit then we can go around it! In the TV films and series Star Trek the starship Enterprise is able to achieve faster than light speed by warping space with its warp drive. Mass and energy warps space. The fabric of space time in Einsteins theory of relativity and the idea you can reconfigure matter creating a shortcut by warping space and making a tunnel. We call them (wormholes), Scientists say to build a wormhole would be relatively straightforward. The problem is they are very unstable, in simple terms they would use negative mass to destabilise regular mass. The universe is made from positive and negative energy in atoms. The idea is to clear a pathos tunnel without getting crushed or exploding in through the process. To enter at one end and exit out the other, there would be no need for interstellar flight and the painstaking lengths of time to traverse the ebony expanse of space. I am not a physicist so would recommend further reading to understand the problem of stability of these Einstein and Rosen bridges commonly know as wormholes! It is believed that black holes may have them but at present there is no proof. In recent years there has been a revolution in particle physics with new and exotic particles being discovered, By building and improving Hadron colliders we are making leaps and bounds. One day I think we might well achieve this possibility. The people that will one day do this will be a radical new us! The journey involves a transcending of us as well as the technology, with power to manipulate the laws of this universe. You may ask why on earth would we want to go through so much trouble to reach this goal when we already have what we want right here? I am happy to repeat, first the fact that Human destruction goes hand in hand with our expansion, population growth and our voracious appetite for natural resources. The second we now know of the great extinctions and courses, so we need a backup plan to guarantee our survival. Thirdly it is in our nature to explore.

CHAPTER 10

Cybermen

First let's start with Robots and AI's for two reasons, the first one is we are already using them and they are evolving fast, we using them in space probes and rovers, just look at a car factory for a good example. The second we can use them to pave the way and do the risky tasks long before we enter an alien environment. Terraforming will be crucial and important Human endeavour. Super-fast computers and AI's Are evolving fast! We already have Home AI assistants which get better every year. One thing that is interesting is they are evolving to be more Human like, for Good or for Bad? We have three possible scenarios, they could be allowed to supersede us and become our masters? I say this because the processing speed of their electronic brains will leave us behind to become the most intelligent. The second we keep them dumbed down, which kind of defeats the object! The third is the one I believe will be the case, that we will be amalgamated with them to create a kind of hybrid. Combining robotics and AI to form a "cyborg"! This may sound scary and even repugnant! Hard line purists may argue that this is not what nature intended or devout religious beliefs that it's not mans job or right to tamper with gods creation, he will provide all what is necessary for us. To answer the concerns I would say nature has always been about change and adapting. If we hadn't tamed fire, we wouldn't have central heating! Whats the point of having our intelligence if we don't use it? More importantly by using it we give ourselves a longer future that is not guaranteed here on earth.

We were once a small mouse like creature, now we are a hairless Ape and tomorrow we will be something else? The God rejection, Well whose God? I mean come on, let's at least get that one sorted. Before we start presuming to understand the mind of God which no one can even agree on which one is the true one! I believe this direction is the natural one for us. We either be the grass blowing in the wind with no control or we can take hold of the reins and steer our own way without being at the mercy of chance and natures whims. The most important fact is we will still be consciousness, it ultimately doesn't matter how you're Mind through leg perceives the ground, flesh or Titanium! Natural evolution is conjectural when it comes to new adaptations? Our organic evolution took a long time, it has taken 6 million years to be us from a chimpanzee type ancestor. We don't have this time anymore and I feel we won't survive long enough to meet the future demands. Our biology will let us down, for the next great transition to be a cosmic race. To start with we will develop AI's and robots to do the really dangerous jobs under our part control. As they evolve even better they will become self-learning and more Autonomous. When it comes to sending these spearhead agents it will be very important they are able to make decisions, because of the time lag between the transmissions back-and-forth from the earth for us to give a command! It takes rovers on Mars 20 minutes to call home with new data, and the further you go the longer the time delay. Having an AI able to make Quick decisions would be crucial to the success of the mission and of course the safety of the device sent there. We could flat pack them with Nano like self assembly habitats and terraforming units plus robots, using memory metals and basic materials for circuits. One such material is carbon tubes which are sheets of graphene, it is super light and strong, and is a hundred times stronger than steel, even tougher than diamonds!

A Brave New world the children of the stars

Armed with the tools and knowledge these robots could start off the evolution of a brand new world, a home for a new civilisation! By the time we get there a lot of the hard work would have already been done. When we do arrive we can supervise and take over the transformation of this brave new world. So when we talk about "Cyborgs and Augmentation" what is it and what would it entail? I just want to make clear that although this chapter is called Cybermen I am also including "Robots and AI's" They will also be used and be independent from Human biology! To clarify when I say Cybermen or cyborg, i'm really saying cyber-Human Referring to Human hybrid technology. This is the route I think we will go down. The other alternative and less attractive is being replaced by the artificial intelligence at some point in the future? When or if it decides we are holding it back and we become redundant in its eyes! I think with millions or billions of years of evolution behind us, and the creators of this technology we have earned the right to survive! The Automatons we will use to assist. Having the ability to make their own decisions with safe guards put in place with minimal Human control. Sent into extreme and dangerous environments to pave the way and one day work beside us and even lookout for us. AI's Have even evolved to a level that can beat the worlds best chess players! It is predicted as we move into the Novasene they will be able to think 10 thousand times faster than us. I hope we can use this new power and not get replaced by it? If AI's become conscious and have some kind of emotions as we do it would be another example of universal consciousness. If not and we go extinct, and say there isn't any other sentient beings in the Universe, what a cold place it would be devoid of the wonder, love and passion for life! The whole point of this trans-humanism is to enhance ourselves and with Bio-technology to live so much longer. We Humans, like all life have been subjected to short lives where nature called the shots and runs it's course and we just had to go along with it. "NO" That just won't do anymore. It's time to take control and be the new law in life, to live one day as immortals or in the short term many thousands of years.

Nature wants 5 of your 7
Children Dead It wants you
Dead by 50 Everything Better
Than that is brought to you by
Science and technology
(David Frum)

Ok what is it to be a Cyborg? Today we are already using bionic limbs, contact lens and cochlear implants. I remember as a child watching the TV series the six million dollar man, the protagonist Steve Austin played by Lee Majors was an American astronaut who had a near fatal crash and is saved by rebuilding him with bionic limbs and an eye. Today this is happening in modern prosthetics, Robots too! Take a look on YouTube and type in "Boston Dynamics" you will be amazed what we can do already. Microchips can be implanted into us as ID integrated devices and linked to an external database, I guess you could use it to unlock your car and front door or even crucial medical information could be stored? In time there will be Bio-chips with super fast processing power that will be thousands of times faster than our normal brains.

A connection to an internet via these micro-chips/Bio-chips would give us a collective link, even in space! Elon Muck CEO of SpaceX is putting over 700 satellites in orbit for internet in space. To see this you can take the Borg as an example in Star Trek to see how it could work, but hopefully not so sinister! People who have claimed to of had encounters with Aliens have reported that when they communicate with them and each other they seem to use some kind of Telepathy. Not all Augmentation has to be mechanical, remember we are organic creatures and there is lot of work that can be done Biologically, DNA manipulation and gene editing are examples. Sending Out Disease seeking Nano cells to correct any emerging problems, also cell replication one of the main reasons for ageing! The main cause of death other than disease, accident death and unfortunately Human Homicide is the ageing process. If you are lucky to Avoid all the possible illnesses and other life threatening causes most of us don't live much past 100 years of age. The reason for this is we are organic and our organs start to fail, we stop successfully renewing our cells. With Bio engineering, using DNA manipulation, Nano medical roaming cells or repair robots and growing replacement organs, like popping into the local garage and getting a new engine! Don't laugh we already do this with "heart transplants" Do you not think about all those people with false teeth in their mouths? We are already slowly becoming hybrids or cyborgs. To start with AI's will probably be ahead of us because crossing Biology with cybernetics is a complex and may I say a moralistic debate among us, especially religious! No one will complain how good or efficient a robot is. However people will not be quite so quick to accept turning into Cybermen! Saying that as the future goes on and we slowly get use to this new technology we will see our lives improve with people walking again, seeing better or even in the dark? Accessing the internet with thought and working in extreme environments it will become accepted and embraced. If God did make man and beast, then he was a bad designer? In our case he put a soft layer of skin, hair, vital blood vessels and organs poorly protected.

If I had designed Man the most important being in Gods eyes on Earth I would of given us some kind of Armoured shell or exoskeleton like beetles, at least a very much tougher skin and maybe the ability to re-grow arms and legs as some lizards do, they can re-grow tails, I have read that even Bacteria in you're gut can make you smarter, making our immune system better and growing healthy brain cells "QED" thus healthier Us, If Human beings are going to embark on being Cosmic people and we want to travel far and live longer we are going to have to re-write the rules. Older generations have always had to stand aside for the "New". We today may think of these metal organic hybrids as abhorrent, but remember those people in the future will be other people evolved and not us. If I lost my leg and God forbid both, I would jump, If you pardon the "Pun" at the chance of the new generation Bionics rather than sitting in a wheel chair. By definition I would be part Cyborg!

The atmospheric pressure of an alien world could be counteracted by a stronger exoskeleton or UV synthetic skin to protect from a more Radioactive parent Star. I'm not sure about Lungs but if the planet is not yet terraformed then there could be some kind of filtering lung depending on how much Augmentation we have and how advanced we become? Like it or not? We will see the rise of the robots hand in hand with Artificial Intelligence to become much more prevalent, we have seen it in the car factories for instance. We even have them mowing the lawn! Children like to play with toy robots and lots of movies depict them as our friends, However there is a perception of them being dangerous? As they get more advanced and without a doubt the AI's Are getting smarter, there may come a day when they become self-aware and turn on us? Just like the film The Terminator! Where skynet decides to get rid of the Humans to become the Dominate intelligence on Earth. Ok it's a good point and yes it could in theory be a fatal turn, What can we do to prevent this? If you go on YouTube there is a video of men attacking a robot which also has a gun and the Robot absolutely will not fight back! Isaac Asimov 1920-1992 was a sci fi writer, professor of Biochemistry and was considered one of the top three science fiction writers! He wrote the famous I-Robot Novel later to become a Hollywood movie staring Will Smith, He wrote in the possibility of robots turning on their Human masters. He then came up with the (Three Laws of Robotics) something that would be programmed into them as the highest prime directive to protect Humans above all else. The rules were introduced in 1942 in a short Runaround which included it in the 1950 collection I-Robot. The Three laws are Quoted from the handbook of Robotics.

The Three laws

. First law

A Robot may not injure a Human being or, through inaction, allow a Human being to come to harm.

. Second law

A Robot must obey the orders given it by Human beings except where such orders would conflict with the first law.

. Third law

A Robot must protect its own existence as long as such protection does not conflict with the first or second law.

I'm not sure what would happen if it meant injuring or killing a Human to save another like preventing a Murder or one life to save Many?

So here is my Forth law

. Forth law

A Robot may not injure a Human being or kill unless justified to protect another Human being.

What we need to do is always have the final say on what we do. If they become sentient? Well, thats a different matter and we would then have to consider some kind of Human rights or AI rights! Being Cyborg we would only really have to think of our own rights. The best bet would be to keep ahead of the subservient Robots! We ourselves will being marrying Micro chips with our brains giving us thousands of time faster thinking and Storing ability. If you are like me? My mind is like a sieve! An onboard updated encyclopaedia, sounds good to me. Along with physical Augmentations to keep up and living so much longer. I hope if one day AI's become sentient that we could co-exist and work together? We only have to look at the Neanderthals to see how one can become extinct. An Ironic scenario is the AI's become our masters and treat us like pets. Mankind has come along way though evolutional time from the days on pond slim! And have earn't the right to be the race we are and to be! As I have said before I Believe we are connected and evolving consciousness no matter what form we take, as long as our spirit is there we have a great future, The future is what we make. Do we as Cybermen have a future in space? YES, unequivocally, the very bodies we will comprise of will be able to look after themselves and indeed be able to interface with the vessels while in Hyperspace while we sleep on our great galactic voyages.

CHAPTER 11

The power of Now

How do we cope with this change previously talked about? This transition into peer machines for instance. Change is nothing new for us, and nine times out of ten doesn't happen out of fear! If you have been on an aeroplane, Do you remember the first time? The feeling of apprehension coupled with existent. This is like facing the unknown, we run the possible "Good and Bad" scenarios through our minds, making that commitment to get onboard. When you leave you're house in the morning, something you have done many times before, you have faith nothing bad is going to happen! It is not a guarantee a meteor won't fall on you're head. Life has always been about chance, just think of the great trails of life on the African grasslands where there are lions looking for their prey. Animals like Impales and Gazelles looking to graze the lands, they know if they go into the open they are in danger of a lion wanting to kill them. The safest thing to do is not do it! Therein lies the problem! If they don't they will starve to death. It is like a way of weighing up the risk, a kind of maybe today I will be ok, We just have to be carful. We do this all the time, conscious or unconscious, it's built in us as "Faith" It isn't exclusive to main stream religion. I am not saying it is perfect, but we as a species are still Here! Anything new is treated with caution in one way or another, it is key to our survival. For us we need to have a clear idea to where we want to go, as a mindful intelligent creature we weigh up the pros and cons. Depending on the situation can sometimes depend on the risks we take?

Life has always been about chance and failings, in fact we exist like all life because more than 99 percent of organisms that have ever lived on Earth are extinct! The universe seems to have no plan, but mutates into being from some cosmic light fuse cord to see what happens? The power of Now is the key to the future, it is the Will of Now that leads us on the path, However as we learn from the Past it doesn't always mean we have to follow in that perceived direction executing our old habits. We have the power to choose! Change begets change, If we want to get into space we have to build a rocket! The future in this realm is an unwritten book, and we don't live in the future only our deep universal consciousness. In this realm it can only be what we want it to be by the "Power of Now" making decisions. The future is the playground of our dreams, the present is like a judge looking back and making comments to the future. It can only be what the Power of Now decides. In the infinite consciousness of the none material world/realm there is only really the state of "Now" By definition time is meaningless in the true conscious realm, You may say future events have happened. I know it is a bit hard to get you're head around that concept, just like quantum physics. I guess what I am saying is don't fear the future, just build it, Yes we will be going through many changes, becoming Cyborgs, but if that makes us better, happier and more life fulfilling than why not? With every new device we invent we soon take it as normal, take the mobile phone, it is metaphorically another limb.

We are now creatures in a Man made time construct. I believe if you see ourselves as a spiritual being experiencing the material world, then you can see being and accepting Augmentation or Cybernetics no more Alien than being organic creatures in physical form! The real revolution is really in the mind and spirit, the thought force and Power behind the physical happenings. In time we will learn to let go of this total individual perception of ourselves. The Power of Now isn't some future mythical state, it is a real Now and present condition, You just have to accept it! By doing this you will finely be able to fully open you're Mind to the endless possibilities. To embrace our deep connectedness in and among all things, not to be afraid of the future. Just be it in the Power of Now! In the universe there is one constant and that is change, I believe in multi realms and dimensions? Without it you wouldn't have Evolution, and in the spiritual realm to create "US" through the process of evolving, exploring and connecting to other states of being. Why do you think our propensity to believe in Gods like we have some deep instinct to be aware of something there, The Now is an instinctual Knowing, we call it intuition! The Form we take as time goes by will evolve with our minds ability and will become familiar with our physical form. Our Minds are not a product of our form, the form is the product of our Minds, and the Mind is the product of consciousness that manifests through. The mind is the gateway from the Now State and becomes a slower vibration, Our minds like all life are like sponges, the bigger the sponge the more water it can hold and like an amplifier is able to produce more of the signal. For us to transform into being Godlike we have to stay the course, when you look at the futuristic Robots and imagine us looking like them it may well seem grotesque, However we can always choose our form! By the choice of the Power of Now we can accept the new Us that will come. By doing this we become different people, we don't have to like it because it won't be us, it will new people, a future people. The Now Me has accepted bionic limbs, cochlear implants, antibiotics and heart transplants. The Power of Now will make and choose what we want to be, grow to be, and what we need! It's all about accepting, deja vu 'accept' it, coincidences are coexisting events just accept it, that feeling of intuition is the deeper you, you guessed it, Accept it! Before I close this short chapter I would like you to at some point in a quiet time and place to look forward and just try to clear you're mind of distraction. I find an open space like a field or park good, but a garden or a room can still work! Take a look and take in what you see and feel it, then put an image of a water surface cutting right through you and everything, but replace the water with the knowledge of the deep universal consciousness and the all encompassing Power of Now, like a flow of energy, then add loads of layers, This may take sometime for some to see it and feel it, but when you do? This is our connection to the other realm.

You have to grow.
From the inside out.
None can teach you,
None can make you
Spiritual. there is no
Other teacher, but
your Own soul

Swami Vivekananda
1863-1902

CHAPTER 12

Transition to be a God

This is the part that becomes more conjectural. We have now moved to look many thousands of years into the future and transformed into something very different to who we are now. The future Us can and should be viewed as always as more enlighten, We today dream up all possible incarnations of extraterrestrial beings in sci-fi with or without robotic creations. We need to keep an open mind of the great steps of achievements we will be able to do in the future? Becoming far more intelligent, living so much longer, if not indefinitely! Achieving enormous amounts of power and occupying the whole Galaxy. Possibly warping through space and time? As I mentioned before in the three types of civilisations, the third one gives us massive amounts of power and even ability to harness the energy of black holes. One proposed idea! One day we might be able to map the Human mind and store a copy of it?

Laser Porting

The conjecture goes that we may one day be able to shed our material bounds and ride on a beam of light. Now this isn't as far-fetched as it first appears! We already use the speed of light with Radio and TV transmissions, how do you think that digital data gets to you're smartphone or WiFi router? Look how we have mapped our DNA and now other Animals and plants. Remember we are conscious energy so half the work has already been done.

The idea is with this digital copy containing our memories, sensations, feelings, personalities and most importantly our spirit! Sent at the speed of light or maybe faster through wormholes, who knows? It would arrive at the planet or place of destination and then be downloaded onto a mainframe or robotic "Avatar" Just imagine the old out of date body You intend replacing, you can simply get stored in a mainframe server then uploaded into the new one. I don't think we would be inserted into a biological body as I have said before in trans-humanism, but into Cyborgs. Although might be possible? I don't see the point! As energy rather than mass will probably use wormholes due to the infinite weight problem of mass, but either way we could catch up with pre-sent missions that paved the way. Our problem today is our bodies let us down, they fail, age and easily get damaged. The Universe is totally held together by energy and life itself is the process of energy, therefore I believe the answer to the Universe is conscious energy, not just in the Universe but outside, adjacent and interlaced, But not bound by it!

Pure Conscious Energy

We came from this energy, consciousness and will ultimately return to who we really are, for we are The Power of Now! Think of you're eyes as the window to you Souls. Imagine you're body is a telescope and the consciousness is looking through it Then it gets knocked over and broken so you can no longer see through it. You are still you, the essence and the soul. There really is no death! Only transformation to another state of being. When you think of speeding to the future in away you are returning to the past and reconnecting to our true selves, we are going home. Whether in a Hybrid coexistence living great lengths of time or short-lived lives the spirit is the same. In fact we are seeing with greater eyes than just ourselves. Being from the Gods, for we are new gods in the making and Gods we shall be. There is an increasing scientific perception that we live in an Holographic universe and something exists because we observe it. What is a Ghost? Answer, perception of energy transformation. Why are they not solid? Answer because they never were! In Genesis in the Bible, In the beginning GOD or GODS started our creation? And this is how I think it should have Been told.

The first day light was created. Mine, The universe came into being by the process of the Big Bang, made by the pure consciousness of the collective gods.

The second day the sky was created, The third day dry land, seas, plants and trees were created. Mine, the laws of physics were created and matter evolves and condensed to form a world environment were life could start.

The forth day the Sun, Moon and the stars were created. Mine, This bit is well out of chronological order it formed before life.

The fifth day creatures that live in the seas and creatures

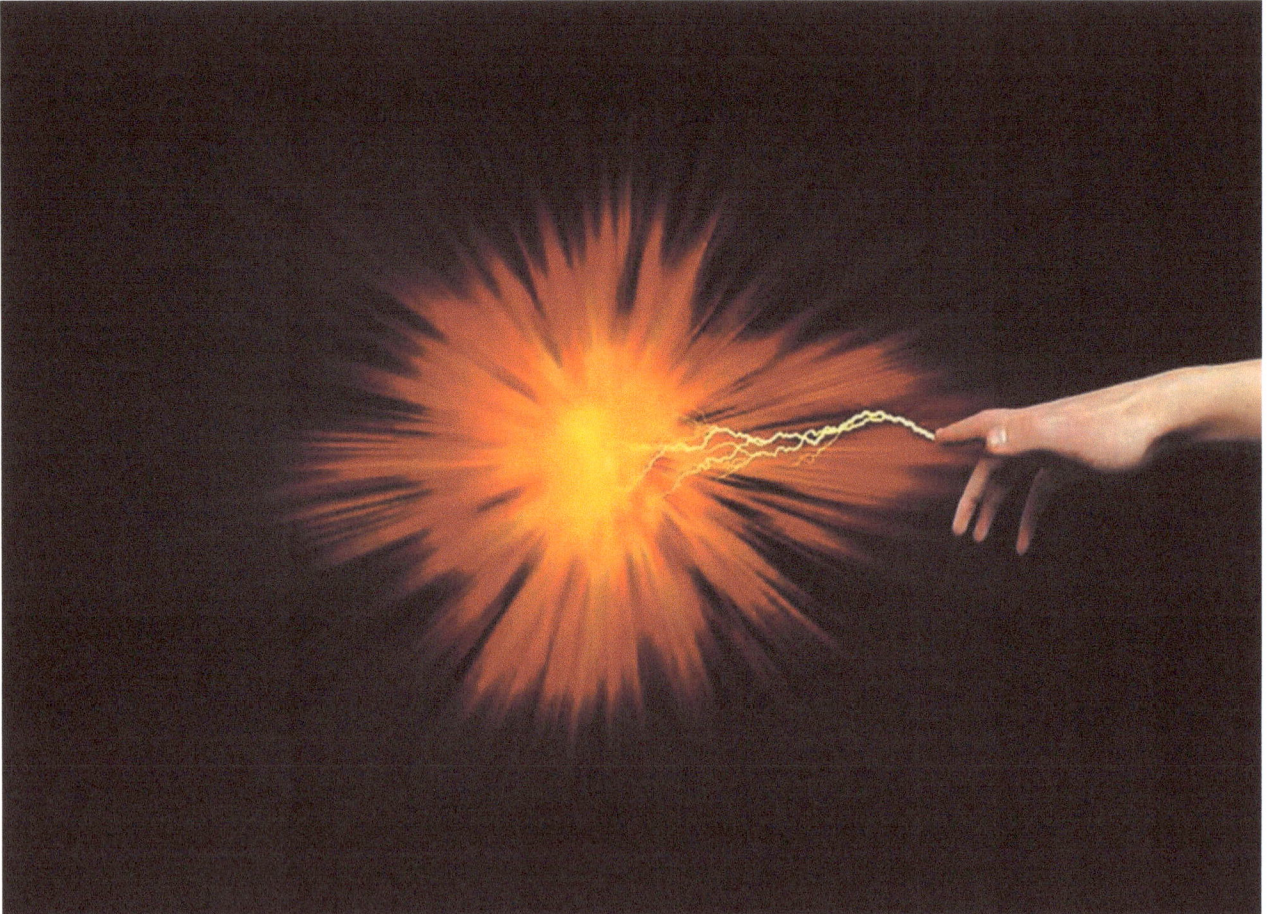

The Gods said let it come into being, and they saw it was good

That could fly were created. Mine, this bit came after the planets were formed and the earth settled down to start the process of life, and then would follow in the right order.

The sixth day animals that live on the land and finally Humans were made in Gods image? I'll leave that to you're imagination! In part correct but Man came about a lot later and animals had been on earth for millions of years prior. What about Gods image? I would say in their imagination? Universal consciousness is thought put into motion and that may or may not have a frame of reference?

The seventh day God finished his work of creation and made it a day of rest and a special day! How does God rest? Or should I say Gods! This is working on the presumption that they work on linear time. As I said in the Paranormal chapter, when people have near death experiences it is reported that it happens seamlessly without a concept of time. Just as atomic particles can be in two different places at the same time, it's an expression of a realm of no time or particular place! Remember time is really only a man made construct, we really only live in a realm of events. Have you ever thought another realm doesn't have to be massive? It Could Be the size of you're brain or infinitely smaller? I believe that we are the creation made up from thought into being, being by pure energy! I Don't know why I call them that? I guess it's because I live in this dimension and don't fully know how to yet relate to pure consciousness, beyond my perception of the normal way we see ourselves. Something I find interesting in the christian belief is it is said that Jesus is the son of God but also is God in Human form in the flesh! The holy trinity, the farther, the son, and the holy spirit. It is said as if they are three different entities, but really only one! Even though I have criticised religion and in this case the christian faith. I think on this statement they were basically right, However not the consciousness of a monotheist deity. I am not sure of how many collective intelligent entities there are? There could in theory be endless amounts? They are universe builders on a spiritual journey.

The journey is in us, A "Conscious journey" our physical bodies are on this journey too! Our spirit has great age and from another place, for it is small at present in Human form, but it will grow and we will evolve into something else, this something will be a "Transition to be a God" We will think of something new and it will come into being! When this time comes we will be back home in the infinite universal consciousness from which we came. A place at a higher level of being, A super vibration at a rate we can't yet imagine. Let me digress for a moment from this ending theme. Let us quickly look at physical evolution, it seems to take a very long time and in the great scheme of things we came along a lot later in the process. When people talk of Visitations from Aliens or more likely inter-dimensional visitors, that seem to come back many years later re-visiting us that is claimed! We wonder why it takes so long do so? Take Jesus again for example, if he was some kind of miracle man maybe he wasn't Human at all? For he is said to return one day? Here we are two thousand years later and no show! Reiterating time is Man made! But for the Gods a re-visit would be instantaneous, for they operate out of time and deep within us so do we. Each realm has its own set of laws and is perceived as such. In this place who knows we may discover that in the big scheme of things we might have more ascending into higher plans or realms, Maybe indefinitely? Planet building, Universe building and cross dimensional travel may just be Childs play. Now we are at the end or the beginning of our spiritual Journey, evolving out of nature and changing our world to the real possibility of being Cosmic people in what ever form we incarnate? This last part is conjecture and my own personal belief, and of course I could be wrong! One thing for sure is we exist and begs an answer! As pre-said it makes more sense for there to be nothing but "Thank the Gods" there is! This is you're journey too, however small you may think it is, but you or we are part of a collective whole, For we are the journey of universal consciousness.

The only end of possible
Is the possibility of not believing
And by believing we are
On a spiritual journey to be
A conscious God
PP

Thanks for reading my conscious journey
And having it with me

This is a conscious and spiritual journey from our Humble beginnings through our evolution. How we perceive ourselves and how we have had to adapt to our changing world. The realisation we have to take care of our home planet, learning how to go to other worlds for our future survival. To take on the great challenges to bridge the great distances to the stars! More importantly how to transform ourselves and our understanding of our place in the Universe. To become Cybermen and be bio-enhanced with the unavoidable destination to becoming "Gods" We will return home whence we came. To understand even as Mortals we are still connected to the Gods, the universal consciousness that is a collective one from our deep past, henceforth the spirit of them is the spirit of us! The Paranormal to some extent is "Normal" it just has to be understood, the unknown is just the undiscovered, more so in ourselves? This book is our journey.

Only those who go so far
Can possibly find how
Far They can go.

Its About
Time

PAUL PHILLIPS

Lightning Source UK Ltd.
Milton Keynes UK
UKHW050508050821
388279UK00002B/39

9 781664 115088